Amino Acids Applications in the Geoenergy Industry

This book discusses the role of amino acids as alternative green additives in the geoenergy and related carbon sequestration industry. It focuses on the role of amino acids in clay swelling inhibition, drilling fluid technology, flow assurance, enhanced oil recovery, gas treatment, geo-carbon separation and sequestration. Amino acids' toxicity, biodegradability, and their properties' effect on managing the various geoenergy operational chemical-related mechanisms are also discussed.

Furthermore, the main challenges involved in the use of amino acids in the geoenergy industries are discussed.

- Discusses mechanisms of relevant amino acids and their possible commercialization in the oil and gas industry
- Addresses properties of amino acids that promise their use in green oil and gas chemistry
- Provides knowledge of the use of amino acids in several aspects of the geoenergy industry
- Clarifies the benefits and drawbacks of using amino acids in the oil and gas industry
- Provides real-world case studies on the use of amino acids in the oil and gas industry

This book is aimed at researchers and graduate students in chemical and petroleum engineering, and geology.

Amino Acids Applications in the Geoenergy Industry

Edited by
Cornelius Borecho Bavoh and
Kwamena Ato Quainoo

CRC Press
Taylor & Francis Group
Boca Raton London New York

CRC Press is an imprint of the
Taylor & Francis Group, an **informa** business

Designed cover image: Getty Images

First edition published 2026
by CRC Press
2385 NW Executive Center Drive, Suite 320, Boca Raton FL 33431

and by CRC Press
4 Park Square, Milton Park, Abingdon, Oxon, OX14 4RN

CRC Press is an imprint of Taylor & Francis Group, LLC

ISBN: 9781032796604 (hbk)
ISBN: 9781032819884 (pbk)
ISBN: 9781003502388 (ebk)

DOI: 10.1201/9781003502388

Typeset in Times
by codeMantra

To:
Janet Adam
Prengel Matema Bavoh Jr.
The Bavoh and Quainoo Families

and

every person (known or unknown)
who believes in the culture of service to humanity
as stewards of the environment.

Contents

Preface

Green chemistry is one of the UN Sustainable Development Goals (SDGs) that are proposed to deal with the replacement of existing hazardous chemicals to ensure a clean environment. Given this, the oil and gas industry has adopted several green chemical development methods/techniques to replace conventional additives. Amino acids have gained much attention as potential green chemicals with wide applications in the oil and gas industry. This book is carefully written to provide a treatise on the use of amino acids as alternative green additives in geo-energy, carbon sequestration, and hydrogen storage within the industry. These include amino acid-based green chemistry applications in the oil and gas industry and carbon capture and storage.

This book specifically focuses on the role of amino acids in clay swelling inhibition, drilling fluid technology, flow assurance (gas hydrates, corrosion, etc.), enhanced oil recovery, gas treatment, geo-storage of hydrogen, geo-carbon separation, and sequestration. Relevant knowledge on amino acids' toxicity, biodegradability, and their properties' effect on managing the various geo-energy operational chemical-related mechanisms is also discussed. Furthermore, the main challenges involved in the use of amino acids in the geo-energy industry are discussed.

The information this book provides would be useful to geoenergy graduate students, academicians, and researchers. In addition, the geoenergy industry and professionals would benefit from the content of this book to solve critical green chemistry issues and meet the SDG goals. Researchers and professionals using green chemistry in geo-carbon sequestration and storage would have a great interest in this book.

Acknowledgments

We thank all the contributors to this edition for their wonderful input and write-ups. Special thanks to Collins for his dedication to compiling this edition. Appreciation to Najat Ibrahim for her proofreading and unwavering support.

Dr. Kwamena Ato Quainoo thanks Sadia Adaliya for her support and companionship during his manuscript preparation.

Finally, many thanks to Janet and Prengel for their support during the writing of this edition.

Editor Biographies

Cornelius Borecho Bavoh is a researcher and academician. He is a lecturer at the University of Mines and Technology, Tarkwa. He worked at the Phase Separation Laboratory of the Research Center for CO_2 Capture ($RCCO_2C$) in the Chemical Engineering Department of Universiti Teknologi PETRONAS (UTP), Seri Iskandar, Perak Darul Ridzuan, Malaysia. He has more than 6 years of experience in tutoring and research at the Universiti Teknologi PETRONAS, Malaysia. Some of his work over the years focus on the development of green solvents (amino acids, biomolecules, deep eutectic solvents, and ionic liquids) for the oil and gas industry. He has worked on several industrial and fundamental projects related to drilling, gas hydrate, and CO_2 capture and separation, and published more than 50 scientific papers in peer-reviewed journals, conference proceedings, and book chapters. Bavoh holds PhD and MSc degrees in Chemical Engineering and BSc (Hons) degree in Petroleum Engineering. His research expertise includes thermodynamics and kinetics of reservoir fluid phase modeling and characterization, gas hydrate in flow assurance, production of naturally deposited methane hydrates, and application of gas hydrate-based technology in CO_2 capture and natural gas storage and transportation, drilling fluid technology, rheology, and cuttings transport. Bavoh is a member of the Society of Petroleum Engineers (SPE).

Kwamena Ato Quainoo is a PhD candidate and a graduate research fellow in the Petroleum Engineering Department at the Missouri University of Science and Technology. He is currently a contributor to an NSF–funded grant to develop a thermo-kinetic model for predicting the precipitation of asphaltene in shale reservoirs. Kwamena received his Master's degree in Petroleum Engineering from the Universiti Teknologi PETRONAS, where he honed his writing skills and developed a deep appreciation for reading cutting-edge scientific papers, writing scientific manuscripts, and conducting peer reviews for scientific journals. To date, he has published thirteen peer-reviewed journal articles and four conference papers, focusing on the use of green chemicals for hydraulic fracturing and drilling fluids, and gas hydrate inhibition including the development of predictive models for asphaltene precipitation and deposition in shale and conventional reservoirs. In addition to his publications, Kwamena regularly reviews manuscripts for top-tier journals such as *Colloid and Polymer Science*, *Energy & Fuels*, *Journal of Dispersion Science and Technology*, *Journal of Rock Mechanics and Geotechnical Engineering*, *Journal of Sol-Gel Science and Technology*, *Journal of Petroleum Exploration and Production Technology*, and *Biomass Conversion and Biorefinery*.

Contributors

Stephen Akuoko is at the University of Mines and Technology, Tarkwa, Ghana.

Eric Mensah Amarfio is at the University of Mines and Technology, Tarkwa, Ghana.

Johannes Ami is at the University of Mines and Technology, Tarkwa, Ghana.

Eric Junior Baffoe is at the University of Mines and Technology, Tarkwa, Ghana.

Cornelius Borecho Bavoh is at the University of Mines and Technology, Tarkwa, Ghana.

Edem Bleppony is at the University of Mines and Technology, Tarkwa, Ghana.

Frank Bonah-Okyere is at the University of Mines and Technology, Tarkwa, Ghana.

Eric Broni-Bediako is at the University of Mines and Technology, Tarkwa, Ghana.

Ato Fanyin-Martin is at the University of Mines and Technology, Tarkwa, Ghana.

Joshua Kojo Aduampong Mantey is at Northwestern University, Evanston, USA.

Solomon Adjei Marfo is at the University of Mines and Technology, Tarkwa, Ghana.

Collins Mensah is at the University of Mines and Technology, Tarkwa, Ghana.

Benjamin Edem Meteku is at the University of Mines and Technology, Tarkwa, Ghana.

Mohammed Abdalla Ayoub Mohammed is at United Arab Emirates University.

Samuel Agyarko Oppong is at the University of Mines and Technology, Tarkwa, Ghana.

Augustine Agyemfra Otu-Koi is at the University of Mines and Technology, Tarkwa, Ghana.

Kwamena Ato Quainoo is at the Missouri University of Science and Technology, Rolla, USA.

Amirun Nissa Rehman is at the King Fahd University of Petroleum and Minerals, Saudi Arabia.

Bennet Nii Tackie-Otoo is at the King Fahd University of Petroleum and Minerals, Saudi Arabia.

Daniel Ekow Yalley is at the University of Mines and Technology, Tarkwa, Ghana.

1 Overview of the Geo-Energy Industry

Cornelius Borecho Bavoh,
Samuel Agyarko Oppong
and Eric Mensah Amarfio

1.1 INTRODUCTION

The world depends on energy to run its activities in the areas of transport, homes, and the business sector. Primary energy sources take the form of nuclear energy, fossil energy (oil, coal, and natural gas), and renewable sources such as wind, solar, geothermal, and hydropower.[1] These primary sources are converted to electricity, a secondary energy source, which flows through power lines and other transmission infrastructure to serve the communities. The share of fossil energy is currently higher than all the other forms of primary energy sources. Due to the increasing warming issues related to fossil fuels, there is increasing demand for the use of other forms of energy. The concept of geo-energy not only encompasses the utilization of domestic subsurface reservoirs as potential energy sources but also as potential storage sites for matter or energy.

The efforts to replace the main or most widely used primary source of geo-energy (fossil fuel energy) with new energy sources cannot be overlooked in this book. However, the focus of this work is on the use of amino acids in the geo-energy industry. Therefore, geo-energy sectors that have received much attention on the application of amino acids will be discussed in this chapter. In view of this, there are active actions to ensure the safe use of fossil energy with less environmental impact. This involves switching from high-carbon energy sources to low-carbon energy sources, a transition mainly propelled by the use of technologically advanced methods for producing fossil fuels with less carbon production. Also, the replacement of toxic additives or chemicals with bio-additives to protect the environment is key to the geo-energy industry. The sustainable development goals aim at reducing carbon emissions by necessitating the use of green chemistry to enhance the production of fossil energy from the earth.[2] The green chemistry applications within the fossil geo-energy sector are primarily applied in drilling, production, and transportation operations.

1.2 THE GEO-ENERGY INDUSTRY

Geo-energy is energy that is stored in the earth, bedrock, groundwater, sediment layers lakes, rivers, or seawater. It can also be defined as naturally occurring energy resources

DOI: 10.1201/9781003502388-1

found within the earth's crust, including oil & gas, oil sands, coalbed methane, and geothermal energy. The geo-energy industry consists of several energy forms produced from the subsurface of the earth. The pioneering geo-energy industry is the fossil fuel energy industry which has served the world for a lot of decades. Although fossil fuel is projected to decline or be replaced with time, it is still an active energy source powering the world, with a highly needed amount.[3] Aside from this geothermal energy is an emerging energy source that is projected to replace fossil fuel energy.[4] Within the geo-energy application, recent technological thematic areas are (1) geothermal technologies, (2) technologies related to capturing, storing, and utilizing CO_2, and (3) technologies related to the unconventional tapping of resources such as shale gas. Among the geo-energy sources, fossil fuel is well established and is currently providing large amounts of energy to the world. This industry consists of several areas that contribute to its success and continued advancements. Specific areas include drilling and fracturing, flow assurance (e.g., corrosion and hydrates), enhanced oil production methods, carbon capture, and hydrogen storage applications in the oil and gas industry. In this book, the various aspects of the geo-energy industry, especially within the fossil fuel area, are discussed. However, below is a brief overview of the geo-energy industry.

1.2.1 Fossil Fuels

Fossil fuel is a natural carbon compound or hydrocarbon that forms in the earth's crust from the remains of prehistoric organisms, including plants, animals, and plankton. This formation occurs within geological environments. Among these compounds, coal, petroleum, and natural gas can be extracted and burned for energy, serving purposes such as powering vehicles or generating electricity through steam turbine generators. All fossil fuels consist of hydrocarbons, which are compounds made up of only two elements: carbon and hydrogen. They originate from the anaerobic decomposition of dead organisms, a process that typically requires millions of years to convert organic material into high-carbon fossil fuels.[5] Because of the lengthy formation time, fossil fuels are categorized as nonrenewable resources. The energy stored in the bonds between hydrogen and carbon atoms is released when these compounds are burned in the presence of oxygen, producing heat energy and carbon dioxide (CO_2). It is anticipated that fossil-based resources will soon become prohibitively expensive, leading humanity to transition to renewable energy sources like solar or wind power. Alternatively, some theorize that carbon-based fuels could be practically limitless when considering all carbon energy sources, such as methane hydrates on the ocean floor, which surpass all other fossil fuel resources combined.[6] Currently, the world continues to depend on fossil fuels such as coal, oil, and natural gas.

1.2.1.1 Coal

Coal is a combustible sedimentary rock that ranges in color from black to brownish-black, forming in geological layers known as coal seams. It is classified as a fossil fuel that develops when decomposed plant matter transforms into peat, which, over millions of years and under immense heat and pressure, becomes coal.[7] Significant coal deposits are derived from ancient wetlands, known as coal forests, which spread across much of the Earth's tropical regions during the late Carboniferous

(Pennsylvanian) and Permian eras. The decay of these plants and ferns, occurring in low-oxygen environments, resulted in coal formation. Primarily made up of carbon, coal also contains varying amounts of hydrogen, sulfur, oxygen, and nitrogen. Its main application is as fuel. Although coal has been utilized for thousands of years, its widespread use only rose during the Industrial Revolution, significantly driven by the development of steam engines. In 2020, coal accounted for approximately 25% of the global primary energy supply and over 33% of electricity generation.[8] Moreover, coal is burned in various industrial processes, including iron and steel production.

The extraction and combustion of coal have detrimental effects on the environment, leading to premature death and illness and constituting the leading human-made source of carbon dioxide, which exacerbates climate change. Coal is categorized into four main types: anthracite, bituminous, sub-bituminous, and lignite. This classification is based on the carbon content and the heat energy potential of the coal, both influenced by the pressure and heat applied to the plants over geological time.

1.2.1.2 Crude Oil

Petroleum is a naturally occurring, yellowish-black liquid made up mostly of hydrocarbons found within geological formations.[9] The term "petroleum" encompasses both unprocessed crude oil and refined petroleum products. Conventional petroleum reserves are primarily extracted by drilling after analyzing the structural geology, sedimentary basins, and characteristics of the petroleum reservoir. In addition, unconventional reserves, such as oil sands and oil shale, are retrieved through methods like fracking. Once extracted, petroleum undergoes refining and is separated typically via distillation into countless products, which can be utilized directly or in manufacturing. These products include fuels like gasoline (petrol), diesel, kerosene, jet fuel, asphalt, and lubricants; chemical reagents for producing plastics; solvents; textiles; refrigerants; paints; synthetic rubber; fertilizers; pesticides; pharmaceuticals; and many more. Petroleum is essential for manufacturing a wide array of materials vital to modern life, with an estimated global consumption of about 16 million cubic meters daily. Its production has significantly influenced industrialization and economic growth.

1.2.1.3 Natural Gas

Natural gas, sometimes referred to as fossil gas, methane gas, or simply gas, is a naturally occurring mixture of gaseous hydrocarbons, with methane constituting approximately 95%, alongside smaller quantities of higher alkanes. It typically contains trace amounts of carbon dioxide, nitrogen, hydrogen sulfide, and helium. Methane is odorless and colorless, being the second-largest greenhouse gas contributor to global climate change after carbon dioxide.[10] To enhance safety, odorants like mercaptan, which has a rotten egg scent, are added to natural gas to facilitate leak detection. The extraction and use of natural gas represent a significant industry. When burned for heat or electricity, natural gas emits fewer toxic air pollutants, less carbon dioxide, and negligible particulate matter compared to other fossil and biomass fuels. During petroleum extraction, natural gas is sometimes flared instead of captured for use. Before natural gas can serve as fuel or be incorporated into manufacturing, it usually requires processing to eliminate impurities such as water. Most natural gas forms through either biogenic or thermogenic processes, with thermogenic gas developing over an extended period when organic material is subjected to heat and pressure deep underground. Methanogenic microorganisms produce methane from various sources, primarily carbon dioxide.

1.2.2 Geothermal Energy

The Earth's crust is composed of massive tectonic plates, which originated from a molten and gaseous state billions of years ago. Beneath the thin crust lies the Earth's core, a highly heated mixture of incandescent material. The interaction between the core and the movement of tectonic plates drives geological phenomena such as hot springs, steam vents, and geysers. The term "geothermal" derives from "geo" (Earth) and "thermal" (heat), signifying the harnessing of Earth's internal heat.[11] Geothermal energy is a renewable, reliable, and efficient energy source, with geothermal power plants capable of operating at capacity factors exceeding 95%. Between 1999 and 2004, global geothermal electricity production grew by 16%, reflecting an annual growth rate of 3%, while direct-use applications increased by 43%, equivalent to an annual growth rate of 7.5%. Geothermal power generation utilizes technologies such as dry steam, flash steam, and binary cycle power plants. By 2010, geothermal electricity was being produced in 26 countries. Geothermal energy systems are categorized into hydrothermal systems, engineered geothermal systems, enhanced geothermal systems, and advanced geothermal systems. Among these, geothermal heat pumps have emerged as one of the fastest-growing renewable energy technologies, leveraging the vast thermal energy stored within the Earth's subsurface. They represent a relatively recent yet well-established innovation in the renewable energy sector.

1.3 GREEN CHEMISTRY IN THE GEO-ENERGY INDUSTRY

The geo-energy industry has long been a basis of the global energy landscape, providing the fuels and chemicals that power modern society. Greenhouse gas emissions and the demand for more energy from the oil and gas industry contribute to climate change and pose significant environmental challenges.[11] Also, the use of environmentally unfriendly additives and chemicals in the geo-energy industry is increasingly leading to pollution. In response, the industry is increasingly adopting eco-friendly strategies to reduce emissions and replace traditionally toxic chemicals. Green chemistry enables the development of advanced products and technologies that help curb emissions from various aspects of the production process of geo-energy.[12]

Green chemistry is a field focused on designing environmentally friendly chemical processes for reducing the use of hazardous chemicals and energy in the geo-energy industry. Green chemistry seeks to replace traditional additives in the industry with more sustainable alternatives. One example is the use of amino acids. This innovation not only reduces the environmental impact but also enhances the selectivity and efficiency of the reactions. Another important aspect of green chemistry is the development of more energy-efficient processes. For instance, supercritical fluid extraction and carbon dioxide capture are green technologies that are being applied to petroleum processing. Supercritical fluids, which exhibit properties between liquids and gases under specific conditions, can be used for efficient extraction of valuable compounds from crude oil, thereby minimizing waste and reducing energy consumption. Similarly, carbon dioxide capture techniques aim to capture and store carbon dioxide emissions from refineries, thus mitigating the industry's contribution to climate change.

The main aim of this book is to discuss the use of amino acids as alternative additives or green chemicals in geo-energy processes. Amino acids are natural compounds composed of amine ($-NH_2$) and carboxylic acid (–COOH) functional groups, linked to the same carbon atom. Different amino acids have a different and specific side chain linked to this exact carbon. There are more than 500 amino acids identified, some of which occur free in nature and some as intermediates in metabolism, but the assembly of proteins within the same species depends on a very small number. Amino acids are environmentally friendly compounds, completely soluble in aqueous media, and can be produced with high purity at low cost and are non-toxic. These properties justify their use as corrosion inhibitors. Also, amino acids consist of amine ($-NH_2$) and carboxylic acid (–COOH) groups which makes them form readily hydrogen bonds with water molecules, and the spontaneous formation of zwitterions enables electrostatic interactions. On the other hand, polar amino acids/peptides (hydrophilic moiety) and nonpolar long-chain compounds (hydrophobic moiety) for building up the amphiphilic structure have high surface activity. They have a low toxicity profile, high biodegradability, and excellent antimicrobial properties against bacteria, fungi, and yeast with high salt and hardness tolerance and are stable at high temperatures. These properties control their application in the geo-energy industry.

1.4 CONCLUSION

In conclusion, the geo-energy industry is a leading energy provider in the world and it is currently subjected to modern technological advances in green chemistry. The use of amino acids exhibits the potential to replace conventional additives used in the geo-energy industry.

REFERENCES

1. Al Mubarak F, Rezaee R, Wood DA. Economic, societal, and environmental impacts of available energy sources: A review. *Eng*. 2024;5(3):1232–1265. doi:10.3390/eng5030067
2. Oliveira JRP, Tusset AM, Andrade DI, Balthazar JM, Pagani RN, Lenzi GG. Action plans study: Principles of green chemistry, sustainable development, and smart cities. *Sustain (Switzerland)*. 2024;16(18):8041. doi:10.3390/su16188041
3. Bagga B. The future of fossil fuels: Renewable energy—wind and solar. In: *Sustainability in the oil and gas sector: Adaptation and mitigation strategies for tackling climate change* (Thomas W, Sergey B, Maya M, Victoria K. Ed.). Switzerland: Cham: Springer Nature; 2024:65-94.
4. Shah M, Prajapati M, Yadav K, and AS. A comprehensive review of geothermal energy storage: Methods and applications. *J Energy Storage*. 2024;98:113019.
5. Himu HA, Raihan A. An overview of integrating the ecological footprints of livestock production with forestry-based strategies. *J Vet Med And Sci Rev*. 2024;1(2):1–13.

6. Zhang X. Methane hydrate development : Geology, technology, and sustainability. *Adv in Resour Res.* 2024;4(1):29–54.
7. Saldaña J, Yurukcu M, Boppana N, Arbabi S, Henry J, Ziyanak S. Conventional and fossil fuels. In: *Energy transition in the oil and gas industry.* 1st ed. [place unknown]: CRC Press Pages42; 2025. p. 42.
8. Bolegenova S, Askarova A, Georgiev A, Nugymanova A, Maximov V, Bolegenova S, Adil'Bayev N. Staged supply of fuel and air to the combustion chamber to reduce emissions of harmful substances. *Energy.* 2024;293:130622.
9. Sai A, Bajpai A, Rohan M, Subburaj B, Subramanian GH. A Novel Approach to Forecasting Crude Oil Price Based on LSTM and Online Learning. *2024 International Conference on Advances in Data Engineering and Intelligent Computing Systems, ADICS 2024.* Chennai, India. 2024:1–6. doi:10.1109/ADICS58448.2024.10533614
10. Climate Council of Australia. Dangerously Overlooked: Why we Need to Talk About Methane. [place unknown]. 2024.
11. Yan M, Livescu S, Dindoruk B, Wisian K. Overview of geothermal systems. In: *Geothermal energy engineering.* (Silviu Livescu and Birol Dindoruk Ed.) [place unknown]: Elsevier; 2025. p. 1–22.
12. Madirisha MM, Dada OR, Ikotun BD. Chemical fundamentals of geopolymers in sustainable construction (Silviu L, Birol D., Ed.). *Mater Today Sustain.* 2024;27(March): 100842. doi:10.1016/j.mtsust.2024.100842

2 Amino Acids Fundamentals

Collins Mensah, Cornelius Borecho Bavoh and Augustine Agyemfra Otu-Koi

2.1 INTRODUCTION

Amino acids are organic molecules containing both a carboxyl (–COOH) and an amine ($-NH_2$) group bonded to the α-carbon atom (the carbon adjacent to the carboxyl group).[1,2] Chemically, each α-amino acid consists of an amino group, carboxyl group, R-group, and hydrogen atom all attached to the α-carbon. While hundreds of amino acids exist in nature or can be synthesized, 20 are considered standard protein-building amino acids. These predominantly occur as L-stereoisomers in biological systems, with D-forms appearing only in specific bacterial products like certain antibiotics and cell wall components. The L- and D-designations describe the absolute configuration of these optically active compounds. All amino acids except glycine have four distinct α-carbon substituents, making them chiral. Eukaryotic proteins exclusively contain L-amino acids. The ionization state of amino acids, which varies with pH, determines their physical properties. Beyond protein synthesis, amino acids serve as precursors for biologically critical molecules including hormones, coenzymes, alkaloids, structural polymers, porphyrins, antibiotics, nucleotides, pigments, and neurotransmitters. Insufficient intake of any single amino acid may disrupt nitrogen balance and impair metabolism.

Comprehending how single amino acids and their combinations influence biochemical and pharmacological pathways is essential for developing new therapeutic compounds. Equally critical is understanding the three-dimensional arrangement of amino acid residues when they bind to receptors or enzymes. Consequently, choosing appropriate amino acids becomes fundamental for maintaining proper biological activity and molecular architecture in drug design.

Alanine (Ala or A), systematically named (S)-2-aminopropanoic acid or α-aminopropionic acid, is a nonpolar, neutral aliphatic amino acid with the chemical structure $HOOCCH(NH_2)CH_3$. This amino acid serves a critical physiological function in the nitrogen transport pathway between skeletal muscle tissue and the liver.[1]

Arginine (Arg or R), systematically named 2-amino-5-(diaminomethylideneamino)-pentanoic acid, is classified as a polar, basic amino acid with a net positive charge at physiological pH. Its chemical structure, $HOOCCH(NH_2)(CH_2)_3NH(C{=}NH_2^+)NH_2$, contains a four-carbon aliphatic backbone terminating in a guanidinium moiety—a planar arrangement of three nitrogen atoms surrounding a central carbon. The conjugated π-system formed between the C=N double bond and nitrogen lone pairs

DOI: 10.1201/9781003502388-2

restricts the guanidinium group to functioning solely as a hydrogen bond donor. Biochemically, arginine serves as the metabolic precursor for nitric oxide (NO), an essential signaling molecule. When incorporated into peptides, arginine residues may facilitate piperidone formation through intramolecular cyclization. Unlike serine (Ser), threonine (Thr), asparagine (Asn), and glutamine (Gln) side chains that participate in bidirectional hydrogen bonding, both arginine and tryptophan (Trp) side chains exclusively donate hydrogen bonds in molecular interactions.[3]

Asparagine (Asn or N), systematically named (2S)-2-amino-3-carbamoylpropanoic acid, is a polar, neutral amino acid with the chemical structure $HOOCCH(NH_2)CH_2CONH_2$. The side chain contains a carboxamide moiety ($-CONH_2$). In sequence notation, Asn may be designated by either Asx or B, symbols that represent ambiguity between asparagine and aspartate. Structurally, Asn residues preferentially occupy terminal positions in α-helical segments and participate in turn conformations within β-sheet secondary structures.[3]

Aspartic acid (Asp or D), systematically named (2S)-2-aminobutanedioic acid, is classified as a polar, acidic amino acid that carries a negative charge under physiological conditions ($HOOCCH(NH_2)CH_2COO^-$). Along with glutamic acid (Glu), its ionized form (aspartate) predominates at biological pH values, accounting for their standard nomenclature as aspartate/glutamate. These dicarboxylate amino acids serve essential catalytic roles as proton donors in enzyme active sites while simultaneously influencing protein characteristics through their effects on aqueous solubility and electrostatic interactions.[3]

Cysteine (Cys or C), systematically named (2R)-2-amino-3-sulfanylpropanoic acid ($HOOCCH(NH_2)CH_2SH$), is a polar, uncharged amino acid containing a thiol (–SH) functional group. Its sulfhydryl moiety confers both hydrophilicity and chemical reactivity, particularly through disulfide bond formation that stabilizes protein tertiary and quaternary structures. The nucleophilic thiol group readily undergoes oxidation to generate cystine (the disulfide-bridged dimer), enabling crucial redox biochemistry. This oxidation-reduction capacity underlies cysteine's antioxidant functions and its role in forming structural crosslinks, exemplified by the essential disulfide bridges in insulin. Furthermore, the thiol group participates in detoxification pathways by conjugating with electrophilic toxins.[3]

Glycine (Gly or G), systematically named aminoacetic acid or aminoethanoic acid ($HOOCCH(NH_2)H$), represents the simplest α-amino acid with nonpolar, neutral aliphatic characteristics. Its minimal side chain (a single hydrogen atom) confers unique structural flexibility, enabling accommodation in sterically constrained regions of protein secondary structures inaccessible to larger residues. Biochemically, glycine serves dual roles as: (1) an inhibitory neurotransmitter in the CNS, modulating neuronal activity; and (2) a metabolic precursor for porphyrin biosynthesis. In protein architecture, Gly collaborates with proline, aspartate, serine, and asparagine residues to stabilize reverse turn conformations. The amino group's acylation potential permits the formation of diacylamide derivatives through sequential acylations.[3]

Histidine (His or H), systematically named 2-amino-3-(1H-imidazol-4-yl)propanoic acid ($HOOCCH(NH_2)CH_2(C_3H_3N_2)$), is a polar, basic amino acid characterized by its imidazole-containing side chain. This heterocyclic moiety exhibits unique acid-base properties, existing in either neutral or protonated states at physiological pH, which accounts for its prevalence in enzyme active sites. The imidazole ring's chemical versatility enables: (1) nucleophilic catalysis when unprotonated (acting as a general base), (2) acid catalysis in its protonated form, and (3) structural stabilization through metal-ion coordination in metalloproteins. Furthermore, the side chain nitrogen's nucleophilicity can mediate intramolecular reactions resulting in both lactam cyclization and proteolytic cleavage of peptide bonds.[1]

Isoleucine (Ile or I), systematically named (2S,3S)-2-amino-3-methylpentanoic acid, is a hydrophobic, aliphatic amino acid with the chemical structure $HOOCCH(NH_2)CH(CH_3)CH_2CH_3$. Classification as a nonpolar, neutral residue derives from its branched hydrocarbon side chain. Among proteinogenic amino acids, isoleucine and threonine are unique in possessing two chiral centers, allowing for four possible stereoisomers. However, biological systems exclusively incorporate the (2S,3S) diastereomer into proteins, reflecting strict enzymatic specificity during biosynthesis.

Leucine (Leu or L), systematically named (S)-2-amino-4-methylpentanoic acid, is a hydrophobic, aliphatic amino acid with the molecular structure $HOOCCH(NH_2)CH_2CH(CH_3)_2$. The branched isobutyl side chain confers its nonpolar character and classification as a neutral amino acid. As one of the three branched-chain amino acids, leucine's aliphatic hydrocarbon moiety strongly influences its role in protein folding and hydrophobic core formation.

Lysine (Lys or K), systematically named 2,6-diaminohexanoic acid ($HOOCCH(NH_2)(CH_2)_4NH_3^+$), is a positively charged, basic amino acid that—along with arginine and histidine—contains a protonatable side chain at physiological pH. The ε-amino group (NH_3^+) exhibits three key biochemical properties: (1) participation in hydrogen bond networks, (2) general base catalysis in enzyme active sites, and (3) formation of covalent crosslinks in structural proteins. These characteristics make lysine indispensable for both collagen biosynthesis and skeletal development through its roles in molecular stabilization and extracellular matrix organization.

Methionine (Met or M), systematically named (S)-2-amino-4-(methylthio)butanoic acid ($HOOCCH(NH_2)CH_2CH_2SCH_3$), is classified as a nonpolar, neutral amino acid containing a thioether moiety. As one of only two sulfur-containing proteinogenic residues (along with cysteine), methionine contributes to cellular redox homeostasis through its antioxidant properties. The metabolically activated form S-adenosylmethionine (SAM) functions as the primary biological methyl group donor in essential transmethylation reactions, including epigenetic modifications and neurotransmitter synthesis.

Phenylalanine (Phe or F), systematically named 2-amino-3-phenylpropanoic acid ($HOOCCH(NH_2)CH_2C_6H_5$), is a hydrophobic, aromatic amino acid characterized by its benzyl side chain. This residue serves dual biological roles: (1) structural contributions to protein architecture through hydrophobic interactions, and (2) biosynthetic precursor for catecholamine neurotransmitters (dopamine, norepinephrine) and thyroid/adrenal hormones. The metabolic disorder phenylketonuria (PKU) results from deficient phenylalanine hydroxylase activity, causing pathological accumulation of phenylalanine. PKU management requires strict dietary restriction of phenylalanine-containing substances, including the dipeptide sweetener aspartame (Asp-Phe-OMe). While chemically stable under most conditions, the aromatic ring undergoes hydrogenation to form cyclohexylalanine derivatives under catalytic conditions.[1]

Proline (abbreviated Pro or P), chemically named (S)-pyrrolidine-2-carboxylic acid, is a nonpolar, neutral amino acid that disrupts regular secondary structures due to its unique cyclic formation. The side chain covalently links back to the nitrogen atom, creating an imino acid with a secondary amine that: (1) cannot donate hydrogen bonds (only accept), (2) forms rigid N-alkyl amide bonds in peptides, and (3) severely restricts backbone flexibility through its pyrrolidine ring. These constrained properties make proline essential for: (a) altering polypeptide chain direction in globular proteins and (b) maintaining structural integrity in collagen-like proteins. The absence of an amide proton and fixed ring geometry account for its classification as a helix-breaking residue.

Serine (Ser or S), systematically named (S)-2-amino-3-hydroxypropanoic acid ($HOOCCH(NH_2)CH_2OH$), is a polar, uncharged amino acid characterized by its hydroxymethyl side chain, representing a hydroxylated alanine derivative. This residue participates in multiple metabolic pathways: (1) nucleotide base biosynthesis (purines/pyrimidines), (2) generation of other proteinogenic amino acids (Gly, Cys, Trp in prokaryotes), and (3) production of complex lipids including sphingolipids. Structurally, Ser contributes to enzyme active sites (e.g., α-chymotrypsin) and, like Asn and aspartate, introduces conformational disruptions in α-helical segments due to its hydrogen-bonding capacity.

Threonine (Thr or T), systematically named (2S,3R)-2-amino-3-hydroxybutanoic acid ($HOOCCH(NH_2)CHOHCH_3$), is a polar, uncharged amino acid classified among the three alcohol-containing proteinogenic residues (with serine and tyrosine). Structurally, it represents a β-hydroxylated valine analog. This amino acid possesses two stereocenters, theoretically yielding four stereoisomers, but biological systems exclusively incorporate the (2S,3R) configuration (L-threonine) into proteins. The uncommon (2S,3S) configuration is designated as L-allo-threonine in biochemical nomenclature.

Serine and threonine exhibit hydrophilic character through their hydroxyl (–OH) functional groups, which participate in hydrogen bond formation. This physicochemical property enables their involvement in various biochemical processes. Notably, threonine has demonstrated specific effects on mammalian feeding behavior, as evidenced by nutritional studies.

Tryptophan (Trp or W), systematically named (S)-2-amino-3-(1H-indol-3-yl) propionic acid (alternatively α-amino-β-indolepropionic acid), is a hydrophobic, aromatic amino acid featuring an indole moiety connected via a methylene bridge. The indole ring system confers pharmaceutical significance, as it appears in numerous bioactive compounds with diverse therapeutic applications including antiemetic, anti-inflammatory, antihypertensive, antimigraine, and antiparkinsonian agents. Biochemically, tryptophan serves as the exclusive biosynthetic precursor for serotonin (5-hydroxytryptamine, 5-HT), a crucial neuromodulator. This amino acid demonstrates particular chemical lability in acidic environments, with its indole side chain susceptible to multiple degradation pathways: (1) oxidative breakdown, (2) dimerization, (3) alkylation, and (4) sulfenyl chloride reactions.

Tyrosine (Tyr or Y), systematically named (S)-2-amino-3-(4-hydroxyphenyl) propanoic acid ($HOOCCH(NH_2)CH_2C_6H_4OH$), is a polar, neutral aromatic amino acid that serves as the metabolic precursor for multiple biologically significant compounds: (1) thyroid hormones (thyroxine), (2) catecholamine neurotransmitters (dopamine, norepinephrine, epinephrine), and (3) the pigment melanin. Impaired tyrosine metabolism can disrupt thyroid hormone production, potentially leading to hypothyroidism. This amino acid exhibits exceptional aqueous solubility, enabling efficient purification from protein hydrolysates. Notably, the post-translationally modified form tyrosine-O-sulfate has been characterized as a constituent of both human fibrinogen and urinary proteins.

Valine (Val or V), systematically named (S)-2-amino-3-methylbutanoic acid ($HOOCCH(NH_2)CH(CH_3)_2$), is a hydrophobic, aliphatic amino acid belonging to the branched-chain amino acid group along with leucine and isoleucine. Its molecular structure features an isopropyl side chain that contributes to several key biological functions including maintaining nitrogen balance, supporting muscle metabolism and repair, and facilitating neuromuscular coordination. The branched nature of valine's side chain creates significant steric hindrance that both decreases the efficiency of coupling reactions in synthetic processes and increases the probability of side reactions occurring. This structural characteristic is shared with isoleucine, another branched-chain amino acid.

Specific amino acids exhibit distinct propensities for influencing protein secondary structure and chemical reactivity. Glutamate, leucine, and methionine preferentially stabilize α-helical conformations, whereas valine, phenylalanine, and isoleucine favor β-sheet formation. Regarding molecular interactions, glycine, proline, valine, and isoleucine demonstrate pronounced cyclization potential. Aspartate and asparagine residues undergo spontaneous conversion to aminosuccinimide derivatives through intramolecular rearrangement. Furthermore, the aromatic side chains of tyrosine and tryptophan, along with methionine's thioether group, show particular vulnerability to alkylation reactions due to their nucleophilic character.

2.2 AMINO ACIDS DEFINITIONS AND STRUCTURE

Amino acids constitute the primary structural components of protoplasmic proteins, characterized by their amphoteric nature resulting from the presence of both amino ($-NH_2$) and carboxyl ($-COOH$) functional groups in their general formula $RCHNH_2COOH$. Glycine represents the structurally simplest amino acid with a hydrogen atom as its R-group (CH_2NH_2COOH), while other amino acids possess more complex side chains that may incorporate additional acidic or basic groups, hydroxyl moieties, sulfur-containing functional groups including thiols ($-CH_2SH$), thioethers ($-CH_2SCH_3$), or disulfide bridges ($-CH_2SSCH_2-$), supplementary nitrogen groups, or cyclic structures.[1] Among the 20 genetically encoded proteinogenic amino acids that predominantly comprise dietary proteins, the common structural framework consists of a central α-carbon atom covalently bonded to four distinct substituents: a hydrogen atom, an amine group, a carboxyl group, and a variable R-group.[4] Proline stands as a structural exception, classified as an imino acid due to its unique cyclic configuration where the side chain forms covalent attachments to both the α-carbon and amine nitrogen, resulting in a secondary amine group rather than the primary amine found in other amino acids.[4] The specific chemical characteristics of the R-group, including its structure and properties, predominantly determine critical physicochemical attributes of the amino acid such as molecular dimensions, net charge, solubility profile, chemical reactivity, and hydrogen bonding potential, which collectively govern the structural and functional properties of the resultant proteins.[5] The D- or L- prefix is omitted when it is obvious that the amino acid is derived from a food protein source thus it is assumed to be an L-amino acids.[6]

The classification of amino acids based on shared characteristics has evolved as a valuable but potentially limiting approach in biochemistry. While these grouping systems provide organizational frameworks, they carry inherent risks of oversimplification. As established in prior discussions, assigning amino acids to rigid categories proves problematic due to context-dependent variability in their functional roles. For example, cysteine participates in distinct biochemical contexts that dramatically alter its replaceability: (1) when forming disulfide bonds, it becomes absolutely unique and irreplaceable; (2) in zinc coordination sites, it may interchange with histidine; while (3) in hydrophobic protein cores, it could be substituted by various nonpolar residues. Additional examples will be examined subsequently.

These classification systems serve primarily as heuristic tools that: (a) organize the 20 proteinogenic amino acids into logical categories, (b) establish clear criteria for identifying amino acid types, and (c) illuminate their characteristic properties. For biochemists, such frameworks significantly improve the understanding of amino acid relationships and distinctions by providing structured approaches to compare and contrast their diverse functionalities.

Amino acids can be systematically categorized through multiple classification schemes that reflect their distinct physicochemical characteristics. One fundamental classification method organizes amino acids according to their isoelectric points, separating them into three major categories: basic, acidic, or neutral. An alternative approach groups amino acids based on their side chain structural features, distinguishing between aromatic and aliphatic types. The most widely used classification system divides amino acids into five principal categories according to their R-group polarity and charge characteristics: nonpolar aliphatic (hydrophobic), aromatic, polar uncharged (hydrophilic), positively charged (basic), and negatively charged (acidic). For greater specificity, Stryer's classification system expands this to seven groups by further differentiating side chain chemistries: aliphatic, hydroxyl-containing aliphatic, aromatic, basic, acidic, amide-containing, and sulfur-containing side chains.

Amino acids can be biochemically classified according to their metabolic fates into glucogenic and ketogenic categories. Glucogenic amino acids are degraded to metabolic intermediates including pyruvate, α-ketoglutarate, succinyl-CoA, fumarate, or oxaloacetate, whereas ketogenic amino acids are converted into ketone bodies. The glucogenic group comprises alanine, cysteine, glycine, serine, threonine, and tryptophan, which are metabolized to either acetyl-CoA or oxaloacetate. Phenylalanine, tyrosine, isoleucine, threonine, and tryptophan exhibit dual metabolic capacity, serving as both glucogenic and ketogenic precursors. In contrast, leucine and lysine are purely ketogenic. Asparagine and aspartate are catabolized exclusively to oxaloacetate. The branched-chain amino acids (leucine, isoleucine, and valine) undergo oxidative metabolism predominantly in extrahepatic tissues including the brain, skeletal muscle, kidneys, and adipose tissue.

2.3 PROPERTIES OF AMINO ACIDS

Amino acids serve as the fundamental structural units of peptides and proteins. Although they share a common structural framework consisting of an amine group, carboxyl group, and variable side chain, the diverse functional groups within these side chains confer unique physicochemical properties to each amino acid that critically determine protein structure and function.[7] This classification system exhibits inherent complexity, as individual amino acids may belong to multiple groups simultaneously. Remarkably, certain amino acids can demonstrate seemingly contradictory properties, with both "hydrophobic" and "hydrophilic" characteristics being applicable depending on specific structural contexts and environmental conditions.

2.3.1 Hydrophobicity of Amino Acids

Hydrophobicity serves as a critical physicochemical parameter for analyzing protein secondary structures, with amino acid classification primarily based on side chain characteristics. Nine amino acids exhibit hydrophobic side chains: glycine (Gly), alanine (Ala), valine (Val), leucine (Leu), isoleucine (Ile), proline (Pro), phenylalanine (Phe), methionine (Met), and tryptophan (Trp). These side chains, composed predominantly of carbon and hydrogen atoms, display minimal dipole moments and demonstrate water-repelling properties that significantly influence protein tertiary structure.[8] Hydrophobic interactions represent fundamental drivers of protein folding and crucial stabilizing forces for various structural configurations, including globular proteins, multiprotein complexes, and protein-ligand interactions. Research indicates that protein hydrophobicity principally originates from amino acid side chains, which constitute the fundamental structural components of protein backbones.[9] The most prevalent classification system divides amino acids into two major categories: hydrophilic (water-preferring) and hydrophobic (water-avoiding). The hydrophobic group can be further subdivided based on side chain structure into aliphatic and aromatic subtypes. This classification reflects their distinct behaviors in aqueous environments and their structural contributions to protein organization.

2.3.2 Aliphatic Side Chains

In strict chemical terms, aliphatic side chains consist exclusively of carbon and hydrogen atoms. By this definition, only five amino acids qualify as truly aliphatic: alanine, isoleucine, leucine, proline, and valine. Alanine's minimal side chain confers relatively weak hydrophobicity, while proline's unique cyclic structure imparts distinct conformational properties that will be examined later. Methionine, despite containing a sulfur atom, is frequently grouped with isoleucine, leucine, and valine due to shared characteristics of flexible, chemically inert side chains well-suited for dense packing in protein cores.

These aliphatic side chains exhibit low chemical reactivity and seldom participate directly in catalytic functions, though they contribute to molecular recognition processes, particularly in binding hydrophobic ligands like lipids. Several additional amino acids contain aliphatic segments within more complex side chains. Arginine, lysine, glutamate, and glutamine demonstrate amphipathic character, combining hydrophobic aliphatic regions (typically two or more –CH2– groups linking the backbone to polar functional groups) with hydrophilic moieties. This dual nature enables these residues to adopt versatile structural roles, with their aliphatic portions potentially buried in hydrophobic protein interiors while polar groups remain solvent-exposed.

2.3.3 Aromatic Side Chains

An aromatic side chain is defined by the presence of a conjugated ring system with delocalized π-electrons, which confers planarity and unique electronic properties. Among the aromatic amino acids, phenylalanine and tryptophan possess strongly hydrophobic side chains, while tyrosine and histidine exhibit intermediate hydrophobicity—often occupying partially buried positions in proteins. Although aromatic residues can sometimes substitute for similarly sized aliphatic residues (e.g., phenylalanine for leucine), larger aromatic side chains like tryptophan cannot functionally replace smaller ones (e.g., valine). Aromatic residues engage in specialized interactions, including π-stacking, where their ring systems align to maximize orbital overlap between delocalized electron clouds.[10] In addition, they mediate specific binding events, particularly with proline-rich sequences. For instance, SH3 and WW domains utilize aromatic residues to recognize and bind polyproline motifs in their interaction partners.[11]

2.3.4 Mutation Matrices

Mutation matrices, also known as substitution or exchange matrices, provide a quantitative framework for classifying amino acids based on their substitution probabilities. These matrices contain numerical values derived from statistical analyses of aligned protein sequences, where the frequencies of observed amino acid substitutions are compared against expected random occurrences. Higher values in the matrix indicate evolutionarily favorable substitutions that occur frequently in nature, while lower values represent less favorable replacements. The calculation of these values typically incorporates evolutionary models to account for varying degrees of phylogenetic divergence between sequence pairs. Two prominent examples of such matrices include the Point Accepted Mutation matrices developed by Dayhoff et al.[12] and the BLOSUM matrices by Henikoff and Henikoff.[13] These tools serve important functions in evaluating the potential effects of amino acid substitutions. Moreover, mutation matrices can be customized for specific analytical needs by incorporating additional biological parameters, such as protein subcellular localization or evolutionary relationships (including both orthologous and paralogous sequences), to better reflect the constraints governing amino acid variability in different contexts.[13] This adaptability enhances their utility in studying protein evolution and function.

2.4 CLASSIFICATION OF AMINO ACIDS

Amino acids are classified based on functional group, polarity and R group, distribution in protein, nutritional requirements, and number of amino and carboxylic groups.[14]

2.4.1 Based on Functional Group

Simple amino acids: these have no functional group in their side chain. Example: glycine, valine, alanine, leucine, and isoleucine.

- Hydroxy amino acids: these have a hydroxyl group in their side chain. Example: serine and threonine.
- Sulfur-containing amino acids: have sulfur in their side chain. Example: cysteine and methionine.
- Aromatic amino acids: have a benzene ring in their side chain. Example: phenylalanine and tyrosine.
- Heterocyclic amino acids: have a side-chain ring that possesses at least one atom other than carbon. Example: tryptophan, histidine, and proline.
- Amine group containing amino acids: derivatives of amino acids in which one of the carboxyl groups has been transformed into an amide group. Example: asparagine and glutamine.
- Branched chain amino acids: a branched-chain amino acid (BCAA) is an amino acid having aliphatic side chains with a branch. Example: leucine, isoleucine, and valine.
- Acidic amino acids: have carboxyl group in their side chain. Example: aspartic and glutamic acid.
- Basic amino acids: contain amino group in their side chain. Example: lysine and arginine.
- Imino acid: amino acids containing a secondary amine group. Example: proline.

2.4.2 Polarity and Function (R) Group

Amino acids with nonpolar R group: these are hydrocarbons in nature, hydrophobic, and have aliphatic and aromatic groups.

Aliphatic R groups: example: alanine, valine, leucine, isoleucine, and proline.

Aromatic groups: example: phenylalanine, tryptophan, and methionine (sulfur).

Amino acids with polar but uncharged R group: these amino acids are polar and possess neutral pH value. Example: glycine, serine, threonine, cysteine, tyrosine, glutamine, and asparagine.

Negatively charged amino acids: their side chain [R group] contains extra carboxyl group with a dissociable proton. And renders electrochemical behavior to proteins. Example: aspartic acid and glutamic acid (Figure 2.1).

Positively charged amino acid: their side chains have extra amino group, rendering basic nature to protein. Example: lysine, arginine, and histidine (Figure 2.2).

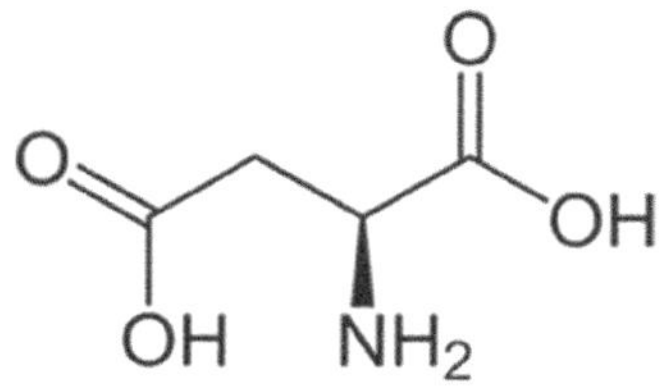

FIGURE 2.1 Structure of aspartic acid.

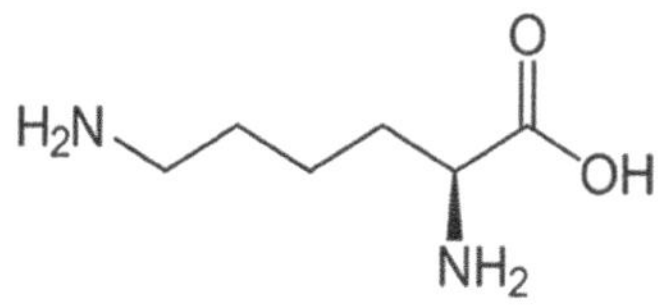

FIGURE 2.2 Structure of lysine.

2.4.3 Distribution in Protein

Standard protein amino acids: the amino acids that are used to form proteins, recognized by ribozyme autoaminoacylation systems. Example: histidine, isoleucine, leucine, lysine, methionine, phenylalanine, threonine, tryptophan, and valine.

Nonstandard protein amino acids: these amino acids are not required to build proteins. These have a vital role as metabolic intermediates. Example: hydroxyproline, hydroxylysine, carboxyglutamate, and diaminopimelate.

Nonstandard nonprotein amino acid: these are the derivatives of amino acids and have role in metabolism. Example: α-amino butyrate, citruline, ornithine, and β-alanine.

2.4.4 Based on Nutritional Requirements

Essential amino acids: essential amino acids cannot be made by the body. As a result, they must come from food. The essential amino acids are: arginine, histidine, isoleucine, leucine, lysine, methionine, phenylalanine, threonine, tryptophan, and valine.

Nonessential amino acids: an amino acid that can be made by humans and so is essential to the human diet. The nonessential amino acids are: alanine, asparagine, aspartic acid, cysteine, glutamic acid, glutamine, glycine, proline, serine, and tyrosine.

2.4.5 On Basis of Number of Amino and Carboxylic Groups

Monoamino monocarboxylic amino acids

- glycine, alanine
- proline

- phenylalanine
- methionine
- serine, threonine

2.5 APPLICATIONS OF AMINO ACIDS

Amino acids have contributed to resolving issues in various fields such as food, infusions, clinical diets, pharmaceuticals, cosmetics, feed, and fertilizer. The functions of amino acids being discovered must realistically contribute to the resolution of issues.

2.5.1 The Food Industry

The chief consumer of amino acids at the present time is the food industry. Problems with enrichment of food products with indispensable amino acids and of balancing the diet (i.e., adding the missing amino acids to low-quality proteins) have been widely discussed in the Soviet and foreign literature, so we will not dwell on them. Food products can be enriched with lysine, threonine (rice, wheat, sorghum, etc.), tryptophan (maize, wheat), and methionine (soybean and peanut flour).[15] However, the balancing of cereal products with pure amino acids (other than lysine) has not yet found widespread application in the food industry, although it improves the nutritional properties of vegetable protein. L-Threonine and L-tryptophan have not yet found practical application for product enrichment, yet vigorous research has been undertaken on the possible use of these important saturated amino acids, which is prevented by their high price. Among possible fields of application of these and other amino acids are baby foods, enrichment of low-calorie food, dietetic feeding, and special diets for cosmonauts. While possessing extremely interesting and diverse properties, amino acids have come to find application in the food industry, as microbiological and chemical methods for obtaining them have developed, mainly as seasonings, to confer taste and aroma on food products, and as antioxidants. The main requirement to be met by various kinds of food additives is that they should not be toxic. Glutamic acid has become the most widely used in the food industry: salt sodium glutamate is commonly employed as an effective taste intensifier. It is widely used to improve the taste of meat and vegetable dishes, as a mixture with table salt, and for modifying the taste of synthetic sweetening agents; in many countries it is added to all products during canning, freezing, or prolonged storage. Glycine is a typical example of sweet amino acids, exhibiting a "refreshing" sweetness equivalent in intensity to that of sucrose. At the present time it is widely used in the Japanese food industry for the manufacture of seasonings, nonalcoholic drinks, soups, marinades, various fish products, and delicatessen. It not only serves as sweetening agent but also creates the foundation of a taste or improves the flavor, by mitigating bitterness and saltiness. It is also used for the modification of artificial products to remove unpleasant aftertaste.

2.5.2 The Pharmaceutical Industry

The low water solubility of some drugs is one of the biggest limitations faced by the pharmaceutical industry, which is associated with problems of oral absorption, formulation, and/or bioavailability (Ambreen et al., 2021; Brunton et al., 2006; Jafari et al., 2021; Kozma et al., 2016). In this context, the combination of drugs with highly water-soluble molecules, such as amino acids, represents a possible alternative to overcome these problems of poorly soluble drugs. Amino acids have proven their capabilities in the pharmaceutical industry in a variety of ways. They can be used alone as nutritional supplements, synthesized to help form other products, or even built into drug delivery systems.[16]

2.6 AMINO ACIDS BIODEGRADABILITY AND TOXICITY

Jones et al. [17] reported that plant roots can directly utilize soil organic nitrogen in the form of amino acids without prior mineralization by the soil's microbial biomass. However, critically assessing this claim requires knowledge of microbial amino acid nitrogen turnover times in soil. A study examined the effects of soil type, depth, and temperature on the uptake and partitioning of a mixture of 15 different 14C-labeled amino acids by the soil microbial biomass across ten contrasting soil types. The results demonstrated that amino acid degradation was soil-dependent, with mean half-lives of 1.7 ± 0.6 hours in topsoils at 18°C compared to 12.2 ± 3.3 hours in subsoils. On average, 34% of the amino acid carbon was respired as CO_2 while 66% was incorporated into new microbial cell biomass. The study also found that amino acid decomposition rates increased with soil temperature. These findings provide important quantitative data on microbial processing of amino acids in different soil environments, which is essential for understanding plant-microbe competition for organic nitrogen sources. The relatively rapid turnover observed, particularly in topsoils, suggests that microbial communities may frequently outcompete plant roots for free amino acids under many soil conditions.[17] Amino acid decomposition in soil is extremely rapid, with typical half-lives for a high concentration of amino acids being in the region of 1 ± 12 h depending on soil type and environmental conditions. As the concentration of amino acids used here is probably around the maximum concentration a rhizosphere soil microorganism might be exposed to, it can be predicted from the concentration-dependent transport kinetics of bacteria and fungi that at lower amino acid concentrations the half-life will be even less.[18] This is supported by evidence from Degens and Harris[19] which suggests that microbial amino acid transporters become saturated at concentrations >10 ± 20 mM. The continued breakdown of proteins by proteases allows the rapid replenishment of amino acids in the soil solution. The rate of amino acid decomposition increased with temperature; however, the specific temperature effect was soil dependent. In all the soils tested, the increased removal of the amino acids from the soil solution was balanced by an increased production of $^{14}C\text{-}CO_2$ biodegradation is the breakdown of chemical compounds by microbes.[17] All the ionic liquids (ILs) were classified as "readily biodegradable" based on their high levels of mineralization (62–87%). The presence of extra carboxyl or amide groups on the amino acid side chain rendered the ILs significantly more susceptible to microbial breakdown. In addition, for most of the [Ch][AA] ILs, low toxicity correlated with good biodegradability. The low toxicity and high biodegradability of these novel [Ch][AA] make them promising candidates for use as environmentally friendly solvents in large-scale applications.[20] Amino acids,

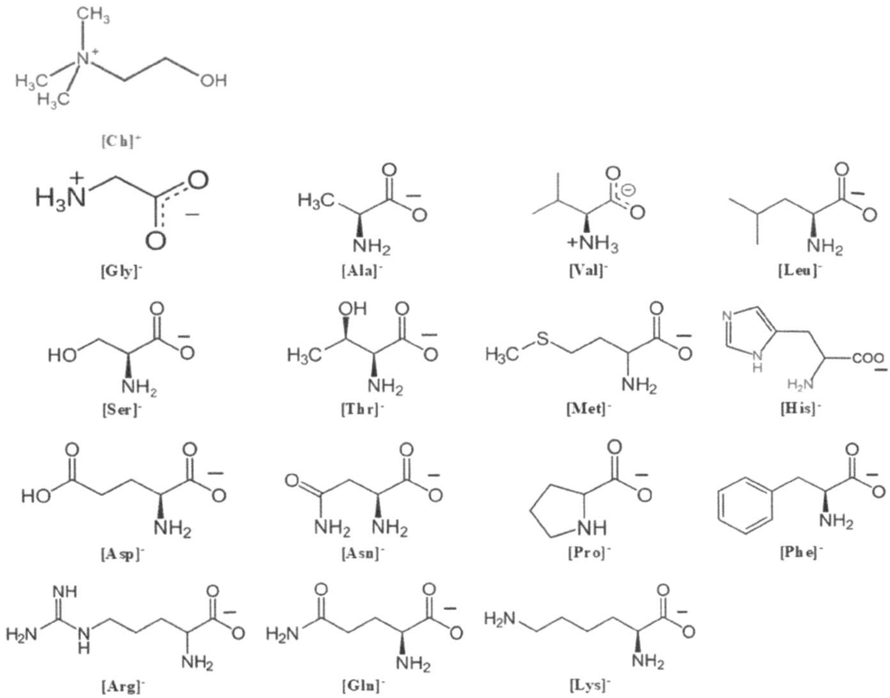

FIGURE 2.3 Structures of the cation and anions of [Ch][AA] ILs.[20]

which have prolific structural diversity as the anionic components of ILs, offer the opportunity for detailed structure-based study of the effect of the anion on IL toxicity and other properties, although to our knowledge toxicity and biodegradation studies of amino acid-based ILs have not been reported previously.[20] Assess their biodegradability using wastewater microorganisms through two different methods—closed bottle and CO_2 headspace tests.[20]

2.6.1 Toxicity toward Bacteria

The study evaluated the antibacterial activity of cholinium-based amino acid ionic liquids ([Ch][AA] ILs) against four representative bacterial strains: *Staphylococcus aureus* (Gram-positive coccus), *Listeria monocytogenes* (Gram-positive rod), *Escherichia coli*, and *Salmonella enteritidis* (Gram-negative rods).[20] The results, expressed as minimum inhibitory concentration (MIC) and minimum bactericidal concentration (MBC), revealed that [Ch][AcO] and [Ch][Cl] exhibited lower toxicity toward the test organisms compared to [Ch][AA] ILs. Contrary to previous enzyme inhibition findings, the imidazolium-based [Bmim][BF4] demonstrated significantly greater toxicity than the cholinium ILs. This observation aligns with prior research showing Gram-positive cocci (particularly *S. aureus*) exhibited greater resistance to various imidazolium and pyridinium ILs.[21] Notably, Gram-negative bacteria generally showed higher susceptibility to the ILs than their Gram-positive counterparts in this study. This finding contradicts the conventional understanding that Gram-negative bacteria, due to their additional outer lipopolysaccharide membrane, typically display greater resistance to biocides compared to Gram-positive bacteria.[22] The results suggest that the antimicrobial mechanism of these cholinium-based ILs may differ from traditional biocides, potentially bypassing or disrupting the protective outer membrane of Gram-negative organisms. These results indicate that the antimicrobial mechanism of these novel ILs likely differs from that of conventional biocides. The significantly lower antibacterial activity of cholinium-based amino acid ILs ([Ch][AA] ILs) compared to imidazolium- and pyridinium-based ILs[23] can be attributed to the inherently low toxicity of the cholinium cation. This conclusion is supported by the observation that the antimicrobial effects of [Ch][AA] ILs were similar to those previously reported for cholinium alkanoate ILs.[24] The study revealed a structure-activity relationship where the toxicity of [Ch][AA] ILs increased with elongation of the amino acid side chain (entries 1–3 and 8, Table 2.1), likely due to enhanced lipophilicity of the anions. This trend mirrors previous findings demonstrating a strong correlation between the carbon chain length of alkanoate anions and IL toxicity. However, the effect was less pronounced when comparing valine to leucine or isoleucine (entries 3–5, Table 2.1), potentially because the branched nature of these amino acid side chains typically reduces their lipophilicity compared to their straight-chain counterparts. A variety of toxicity effects were given by the inclusion of polar groups, such as hydroxyl, carboxyl, and amino, into the anions, as indicated in entries 6, 7, and 9–15 (Table 2.1). The amino acid's antibacterial activity increased slightly when a hydroxyl group was added to its side chain (entry 2 vs. item 6, Table 2.1). Conversely, the carboxyl group provided little toxicity on entries 9 and 10 of Table 2.1 for [Ch][Asp] and [Ch][Glu], which, out of the [Ch][AA] ILs examined, are the least hazardous. The [Ch][Asn] and [Ch][Gln], amide analogs, were somewhat more hazardous (Table 2.1, entries 11 and 12). The amino acid side of the ILs chains with fundamental functional groups showed significant toxicity (Table 2.1, entries 13–15), which was similar to the toxicity of [Ch][Val]. The significant toxicity differential between [Ch][AA] having side chains of acidic and basic amino acids may be partially accounted for by the ILs' alkalinity (entries 9–10 vs. entries 13–15, Table 2.1). The pH values of the [Ch][AA] ILs containing basic amino acids and anions in a 5 mM solution were much higher (10.0–11.3) than those of identical solutions containing ILs containing acidic amino acids, which had pH values of about 7.[25] These ILs based on basic amino acids would be much more prevalent in the environment at high concentrations, which would make microorganisms more susceptible to harm. The results of the tests indicated that [Ch][Phe] and [Ch][Trp] were the ILs that were most toxic to bacterial cells overall (entries 17 and 18, Table 2.1), indicating that the ILs' toxicity was significantly increased when aromatic systems were added to the anions.

2.6.2 Biodegradability

The closed bottle and CO_2 headspace tests, in which [Ch][AA] ILs were added to aerobic aqueous media inoculated with wastewater microorganisms, and the

TABLE 2.1
Toxicity of [Ch][AA] ILs toward Bacteria

IL	MIC	MBC	MIC	MBC	MIC	MBC	MIC	MBC
[Ch][Gly]	125	250	93.8	125	250	500	250	500
[Ch][Ala]	46.9	93.8	62.5	93.8	188	250	188	250
[Ch][Val]	46.9	62.5	46.9	62.5	93.8	125	93.8	125
[Ch][Leu]	46.9	62.5	46.9	62.5	46.9	62.5	93.8	125
[Ch][Ile]	46.9	62.5	46.9	62.5	46.9	62.5	93.8	125
[Ch][Ser]	31.3	62.5	46.9	62.5	93.8	125	125	250
[Ch][Thr]	31.3	62.5	46.9	62.5	93.8	125	125	250
[Ch][Met]	31.3	62.5	31.3	62.5	25	250	125	250
[Ch][Asp]	500	750	125	188	500	750	750	750
[Ch][Glu]	500	750	125	188	500	750	750	750
[Ch][Asn]	93.8	125	93.8	125	250	250	188	250
[Ch][Gln]	125	188	93.8	125	188	250	125	188
[Ch][Lys]	31.3	46.9	31.3	48.7	62.5	93.8	62.5	93.8
[Ch][His]	46.9	62.5	46.9	62.5	93.8	125	93.8	125
[Ch][Arg]	46.9	93.8	46.9	62.5	93.8	125	93.8	125
[Ch][Pro]	46.9	62.5	46.9	62.5	62.5	93.8	93.8	125
[Ch][Phe]	31.3	46.9	31.3	46.9	46.9	62.5	62.5	93.8
[Ch][Trp]	23.4	31.3	23.4	31.3	31.3	46.9	46.9	62.5
[Ch][AcO]	500	750	750	750	500	750	500	750

depletion of dissolved O_2 or the evolution of CO_2 was periodically determined, were the two standard methods used to evaluate the aerobic biodegradability of the [Ch][AA] ILs.[20] The reference chemical sodium benzoate and [Ch][AcO] were used to examine the biodegradability of the [Ch][AA] ILs (Tables 3 and S2, accessible in File S1). In percentage terms, the biodegradability was stated relative to the theoretical maximum. In general, if the reference compound of 60% biodegrades in 14 days, the biodegradation test procedures are deemed legitimate. The biodegradability of sodium benzoate in this investigation was determined to be .75% in 14 days using both approaches, proving the reliability of the two test techniques. The biodegradation test is said to be passed by compounds that, after 28 days, achieve a biodegradation level of .60%, according to the Organization for Economic Cooperation and Development (OECD).[23] All of these ILs can be regarded as "readily biodegradable" because their biodegradability after 28 days was greater than 60% according to CO_2 headspace tests and the closed bottle (Table 2.2). Compared to [Ch][AcO], which was previously found to be a low-toxicity IL,[26] the majority of [Ch][AA] ILs shows greater susceptibility to aerobic biodegradation. The cholinium cation and the presence of carboxyl and primary amino groups in the anions may be responsible for the high levels of biodegradation of the ILs.[27] A possible explanation for the variations in the aerobic biodegradability of different [Ch][AA] ILs is the structure of amino acid anions. In comparison to the >80% biodegradation observed with the unbranched side chains of [Ch][Gly] and [Ch][Ala] (entries 1 and 2, Table 2.2), the biodegradability of 69–72% (entry 3–5, Table 2.2) was achieved

TABLE 2.2
Biodegradation of ILs Determined by the Closed Bottle Test

Entry	ILs	7 days	14 days	21 days	28 days
1	[Ch][Gly]	58.3 ± 2.3	72.3 ± 0.1	75.0 ± 1.2	82.6 ± 1.1
2	[Ch][Ala]	61.9 ± 0.8	66.8 ± 0.1	77.6 ± 1.2	80.0 ± 0.4
3	[Ch][Val]	49.6 ± 0.1	61.3 ± 0.8	65.6 ± 0.5	69.4 ± 0.6
4	[Ch][Leu]	46.3 ± 0.8	68.5 ± 1.6	70.7 ± 0.9	72.4 ± 0.1
5	[Ch][Ile]	57.3 ± 2.4	68.4 ± 0.1	70.5 ± 0.3	71.6 ± 0.8
6	[Ch][Ser]	53.8 ± 2.2	72.0 ± 0.7	73.5 ± 1.7	80.6 ± 1.5
7	[Ch][Thr]	44.6 ± 2.3	64.9 ± 1.7	70.0 ± 1.1	74.3 ± 1.5
8	[Ch][Met]	54.3 ± 0.3	63.7 ± 0.1	64.7 ± 0.3	66.1 ± 0.9
9	[Ch][Asp]	68.9 ± 1.2	79.5 ± 0.6	86.5 ± 0.5	87.1 ± 0.6
10	[Ch][Glu]	70.0 ± 0.1	71.3 ± 1.1	83.1 ± 1.4	86.3 ± 0.2
11	[Ch][Asn]	53.9 ± 0.5	71.8 ± 1.1	85.7 ± 0.7	87.1 ± 1.2
12	[Ch][Gln]	58.5 ± 2.3	80.5 ± 0.7	83.8 ± 1.5	86.6 ± 1.4
13	[Ch][Lys]	54.4 ± 3.7	62.4 ± 0.4	65.9 ± 1.0	67.7 ± 1.0
14	[Ch][His]	46.5 ± 3.5	60.1 ± 1.2	63.4 ± 0.0	65.3 ± 1.3
15	[Ch][Arg]	59.6 ± 2.3	62.3 ± 0.9	65.3 ± 0.1	67.6 ± 0.2
16	[Ch][Pro]	58.0 ± 0.9	66.4 ± 0.1	68.5 ± 0.9	71.3 ± 0.1
17	[Ch][Phe]	44.0 ± 0.2	68.8 ± 0.2	71.0 ± 1.1	70.8 ± 0.1
18	[Ch][Trp]	55.1 ± 1.1	60.2 ± 0.6	62.7 ± 1.1	65.9 ± 0.2
19	[Ch][AcO]	46.7 ± 0.4	63.6 ± 0.8	66.5 ± 0.5	68.1 ± 1.9
20	Sodium benzoate	60.1 ± 0.2	75.0 ± 1.0	78.5 ± 0.6	81.1 ± 0.8

through the use of amino acids with branched side chains, such as Val, Leu, and Ile, as the anions. This boosted resistance to aerobic biological breakdown by microorganisms. With the unbranched side chains of [Ch][Gly] and [Ch][Ala], 80% biodegradation was seen (entries 1 and 2, Table 2.2). Likewise, Petkovic et al. found that branched chain alkanoate-containing ILs.

Anions resisted fungal invasion better than their linear isomers.[26] Due to the hydroxyl in the anions, there was a small An enhancement in biodegradation was also noted in ILs of other kinds.[28] For example, as a result of the substitution of a hydroxyl group for a methyl group, the degree of the biodegradation of [Ch][Thr] was somewhat greater than [Ch][Val] (entry 7 vs. entry 3, Table 2.2; 74% vs. 69%). The ILs adding amide and carboxylic acid groups demonstrated outstanding degradability (.86%, Table 2.2, entries 9–12). These findings are consistent with Boethling's general guidelines, which state that the presence of carboxylate groups or hydrolyzable bonds, such as amide and ester, generally increases aerobic biodegradability.[27] However, they differ slightly from a study by Gathergood et al., which found that ILs incorporating an amide group in the cation side chain showed negligible biodegradation.[29] Degradation values of 65–68% were observed for the ILs containing basic amino acids, indicating that they were less vulnerable to microbial breakdown (entries 13–15, Table 2.2). Since these ILs' concentrations in the test systems were incredibly low (3 mg L21), it is doubtful that their poor biodegradation was caused by their significant basicity. The heterocyclic imidazole group, which may be detrimental to biodegradation, may be the cause of [Ch][His]'s comparatively low biodegradability.[27] Comparably, [Ch][Trp] with its heterocyclic indole double ring structure had a somewhat lower biodegradability (71% vs. 66%, entries 16 and 17, Table 2.2).[27] Similarly, [Ch][Phe] with its single aromatic carbon ring structure had a slightly higher biodegradability. The biodegradability test was previously reported to be unsuccessful for cholinium 2-naphthoxyacetate and cholinium anthracene-9-carboxylate, both of which contain polycyclic aromatic hydrocarbon ring moieties [16].

The CO_2 headspace test showed a similar pattern, indicating that the structures of amino acid anions affect the biodegradability of ILs. For instance, the ILs with amide and carboxyl groups have shown exceptional biodegradability. Moderate biodegradation (63–66%) was seen in the cases of the ILs with basic amino acid anions. Similarly, it seems that [Ch][Trp] was resistant to microbial assault, allowing for a 62% biodegradation level. A tension between optimizing biodegradability and limiting toxicity has occasionally been seen in the rational design of ecologically benign chemicals.[30] For instance, there was a correlation between increased toxicity and side chain elongation that led to improved biodegradation of ILs based on pyridinium.[31] As shown in Figure 2.3, the toxicity of [Ch][AA] ILs against AChE and *E. coli* was compared with the biodegradability data in order to examine the relationship between the two characteristics. It's interesting that these [Ch][AA] ILs did not exhibit this kind of contradiction between toxicity and biodegradability. With comparatively few exceptions, the degradability of the [Ch][AA] ILs with lower toxicity was higher. As an illustration, the easily biodegradable [Ch][Asp] and [Ch][Glu] showed low toxicity, while the more poisonous [Ch][Trp] was discovered to be more resistant to microbial degradation.

REFERENCES

1. Milne PJ, Kilian G. The properties, formation, and biological activity of 2,5-diketopiperazines. *Comprehensive Natural Products II: Chemistry and Biology*. 2010;5:657–698. doi:10.1016/b978-008045382-8.00716-4
2. Maloy S. 2013. *Amino Acids*. [place unknown]: Elsevier Inc. doi:10.1016/B978-0-12-374984-0.00051-6
3. Nelson DL, Cox MC, Freeman WH. 2004. Lehninger Principles of Biochemistry. 4th ed. [place unknown].
4. Ustunol Z. 2015. 3 Amino Acids, Peptides, and Proteins. [place unknown].
5. Nelson DL, Cox MM. Lehninger Principles of Biochemistry, 4th Edition. [place unknown].
6. Ustunol Z. 2008. Physical, Chemical, and Processing-Induced Changes in Proteins.
7. Anon. 2020. Amino Acid Physical Properties Table of amino acid physical properties [Internet]. [accessed 2024 Oct 31]:1–2. https://www.thermofisher.com/hk/en/home/life-science/protein-biology/protein-biology-learning-center/protein-biology-resource-library/pierce-protein-methods/amino-acid-physical-properties.html
8. Anon. Biomolecules : Protein 1 Hydrophobic Amino Acids Polar Amino Acids [Internet]. :2. https://www2.chem.wisc.edu/deptfiles/genchem/netorial/modules/biomolecules/modules/protein1/prot13.htm
9. Zhu C, Gao Y, Li H, Meng S, Li L, Francisco JS, Zeng XC. Characterizing hydrophobicity of amino acid side chains in a protein environment via measuring contact angle of a water nanodroplet on planarpeptide network. *Proc Natl Acad Sci U S A*. 2016;113(46): 12946–12951. doi:10.1073/pnas.1616138113

10. Hunter CA, Singh J, Thornton JM. Π-Π Interactions: the geometry and energetics of Phenylalanine-Phenylalanine interactions in proteins. *J Mol Biol*. 1991;218(4):837–846. doi:10.1016/0022-2836(91)90271-7
11. Macias MJ, Wiesner S, Sudol M. WW and SH3 domains, two different scaffolds to recognize proline-rich ligands. *FEBS Lett*. 2002;513(1):30–37. doi:10.1016/S0014-5793(01)03290-2
12. Dayhoff MO, Hersh RT, Eck R V. Atlas of protein sequence and structure, 1966. *Syst Zool*. 1967;16(3):262. doi:10.2307/2412074
13. Henikoff S, Henikoff JG. Amino acid substitution matrices from protein blocks. *Proc Natl Acad Sci U S A*. 1992;89(22):10915–10919. doi:10.1073/pnas.89.22.10915
14. Amino Acids Classification and Properties.
15. Stopka K. Industrial applications of Ozone. *Ind Wastes*. 1978;24(3):23–24.
16. Anon.TheRoleofAminoAcidsinPharmaceuticalsWorkwithOakwoodLabsonYourNext Project Learn More About Using Amino Acids in Pharmaceuticals [Internet]. [accessed 2024 Nov 1]. https://oakwoodlabs.com/the-role-of-amino-acids-in-pharmaceuticals/
17. Jones DL. Amino acid biodegradation and its potential effects on organic nitrogen capture by plants. *Soil Biol Biochem*. 1999;31(4):613–622. doi:10.1016/S0038-0717(98)00167-9
18. Amino Acid Transport and Pool Formation in Pseudomonas Aeruginosa by B. S. A. (Microbiology), University of British Columbia, 1963 M. Sc. (Microbiology), University of British Columbia, 1966A Thesis Submitted I. 1968.
19. Degens BP, Harris JA. Development of a physiological approach to measuring the catabolic diversity of soil microbial communities. *Soil Biol Biochem*. 1997;29(9–10):1309–1320. doi:10.1016/S0038-0717(97)00076-X
20. Hou XD, Liu QP, Smith TJ, Li N, Zong MH. Evaluation of toxicity and Biodegradability of Cholinium Amino Acids Ionic Liquids. *PLoS One*. 2013;8(3):e59145. doi:10.1371/journal.pone.0059145
21. Docherty KM, Kulpa CF. Toxicity and antimicrobial activity of imidazolium and pyridinium ionic liquids. *Green Chem*. 2005;7(4):185–189. doi:10.1039/b419172b
22. Russell AD. Similarities and differences in the responses of microorganisms to biocides. *J Antimicrob Chemother*. 2003;52(5):750–763. doi:10.1093/jac/dkg422
23. Wu S, Li F, Zeng L, Wang C, Yang Y, Tan Z. Assessment of the toxicity and biodegradation of amino acid-based ionic liquids. *RSC Adv*. 2019;9(18):10100–10108. doi:10.1039/c8ra06929h
24. Petkovic M, Ferguson JL, Gunaratne HQN, Ferreira R, Leitão MC, Seddon KR, Rebelo LPN, Pereira CS. Novel biocompatible cholinium-based ionic liquids—toxicity and biodegradability. *Green Chem*. 2010;12(4):643–664. doi:10.1039/b922247b
25. Liu QP, Hou XD, Li N, Zong MH. Ionic liquids from renewable biomaterials: synthesis, characterization and application in the pretreatment of biomass. *Green Chem*. 2012;14(2):304–307. doi:10.1039/c2gc16128a
26. Boethling RS, Sommer E, DiFiore D. 2007. Designing small molecules for biodegradability. *Chem Rev*. 107(6):2207–2227. doi:10.1021/cr050952t
27. Ford L, Harjani JR, Atefi F, Garcia MT, Singer RD, Scammells PJ. Further studies on the biodegradation of ionic liquids. *Green Chem*. 2010;12(10):1783–1789. doi:10.1039/c0gc00082e
28. Gathergood N, Garcia MT, Scammells PJ. Biodegradable ionic liquids: part I. Concept, preliminary targets and evaluation. *Green Chem*. 2004;6(3):166–175. doi:10.1039/b315270g
29. Wu S, Li F, Zeng L, Wang C, Yang Y, Tan Z. Assessment of the toxicity and biodegradation of amino acid-based ionic liquids. *RSC Adv*. 2019;9(18):10100–10108. doi:10.1039/c8ra06929h
30. Docherty KM, Dixon JK, Kulpa CF. Biodegradability of imidazolium and pyridinium ionic liquids by an activated sludge microbial community. *Biodegradation*. 2007;18(4):481–493. doi:10.1007/s10532-006-9081-7

3 Amino Acids as Gas Hydrate Inhibitors

Collins Mensah, Edem Bleppony and Cornelius Borecho Bavoh

3.1 INTRODUCTION

Gas hydrates are ice-like crystalline compounds formed by the trapping of gas molecules within hydrogen-bonded water molecules under high-pressure and low-temperature conditions. The gas molecules are trapped in the water molecules through van der Waals forces.[1] Depending on the type, shape, and size of the gas molecules, three basic gas hydrate structures can occur: cubic structure I, cubic structure II, and hexagonal structure H. Figure 3.1 illustrates the available gas hydrate structures.[2] Gas hydrates have various applications, such as serving as a future energy source,[3] facilitating CO_2 capture and gas separation,[4,5] and enabling the storage and transportation of gases, including natural gas, hydrogen, and carbon dioxide.[6–8]

Conversely, gas hydrates pose significant flow assurance challenges in the oil and gas industry. During hydrocarbon drilling, production, and processing operations, gas hydrates can form in pipelines and facilities, leading to blockages, substantial costs for prevention and removal, environmental hazards, and sometimes loss of life.[1] Techniques such as heating, water removal, depressurization, and chemical injection are used to prevent or remove gas hydrate plugs in pipelines. Among these, chemical injection is widely employed due to its economic and technological feasibility.[1,10] Generally, two major types of gas hydrate chemical additives (inhibitors or promoters) are used, depending on the area of application. These additives influence gas hydrate formation either thermodynamically, by altering the hydrate phase equilibrium boundary conditions, or kinetically, by enhancing or delaying the nucleation and crystal growth rate of hydrate formation.

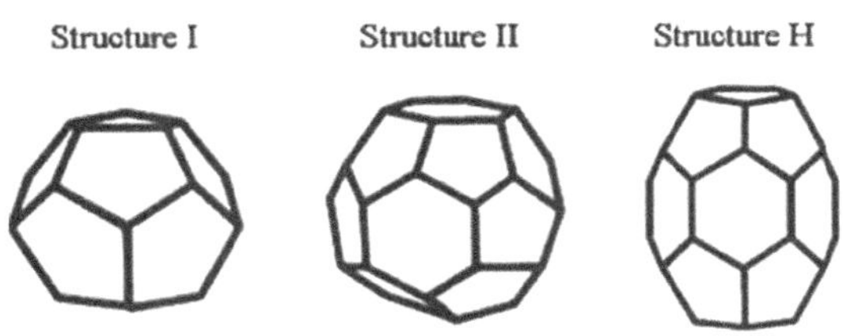

FIGURE 3.1 Common gas hydrate crystal structures.[9] Copyright © 2014, American Chemical Society.

DOI: 10.1201/9781003502388-3

Thermodynamic hydrate inhibitors (THIs) and low-dosage hydrate inhibitors (LDHIs) are the available chemical inhibitors. THIs, such as glycols and methanol, inhibit gas hydrates thermodynamically by reducing the activity of water in hydrate formation through the formation of hydrogen bonds with water molecules. Consequently, they expand the non-hydrate formation region of the hydrate formation phase boundary by shifting the equilibrium hydrate formation curve to higher pressures and/or lower temperatures. The use of THIs requires high concentrations, leading to high operational costs. At high subcooling temperatures, concentrations over 40 wt% are often required to ensure inhibition. Moreover, THIs are highly volatile and environmentally prohibited. Alternatively, LDHIs consist of kinetic hydrate inhibitors (KHIs) and anti-agglomerants. KHIs are typically polymers, such as polyvinylpyrrolidone (PVP) and poly(N-vinylcaprolactam), and they prevent gas hydrate formation by adhering to hydrate crystals to prolong or delay hydrate nucleation time (induction time) and growth rate. Although KHIs are used at low concentrations (<2 wt%), they are ineffective at high subcooling and shutdown conditions. Therefore, it is crucial to introduce new chemical inhibitors that are environmentally friendly, less expensive, and highly effective in addressing these issues.[11]

The application of hydrate-based technology for carbon capture and sequestration (CCS) and natural gas storage involves the use of chemicals to enhance hydrate formation, in contrast to the prevention of hydrates in flow assurance systems. Gas hydrate-based CCS initially involves a CO_2 separation process via the formation of CO_2 hydrates in a CO_2 mixed gas system (e.g., flue gas and natural gas). Since CO_2 is highly prone to hydrate formation at low pressures, it forms hydrates more rapidly with a higher gas (CO_2) to hydrate conversion ratio than other gases. The residual gas can be transferred to a vessel, as illustrated in Figure 3.2. The rich CO_2 hydrates are then dissociated to extract CO_2 for further sequestration processes, similar to hydrate-based natural gas storage processes. The separated CO_2 can be sequestered or stored in reservoirs in hydrate form. In addition, CO_2 hydrates can be deposited as hydrate pellets on the seabed, provided they remain stable. Thermodynamic hydrate promoters (THPs) and kinetic hydrate promoters (KHPs) are the available gas hydrate chemical promoters. THPs are primarily used to shift the hydrate phase boundary conditions to higher temperatures and lower pressure regions. KHPs, on the other hand, are employed to increase the hydrate induction time, formation rate, and the gas/water uptake during hydrate formation. Commonly used THPs include tetrahydrofuran (THF)[12] and acetone, while nanoparticles,[13] sodium dodecyl sulfate (SDS),[14,15] and some other surfactants serve as KHPs. THPs and KHPs are applied in CO_2 capture and sequestration[5,16] as well as in gas storage and transportation.[17,18] However, these conventional promoters, like conventional inhibitors, are environmentally prohibitive and less effective.

Based on the general knowledge that compounds exhibiting strong electrostatic charges and/or a strong affinity for hydrogen bond formation can inhibit gas hydrate formation, novel gas hydrate inhibitors have been introduced to potentially replace existing commercial inhibitors. One such class of inhibitors is ionic liquids (ILs),[10,19–21] which have garnered attention due to their zero volatility and dual functionality in hydrate inhibition,[21] acting as both THIs and KHIs. More detailed information on ILs as gas hydrate inhibitors can be found in references [10, 22–24].

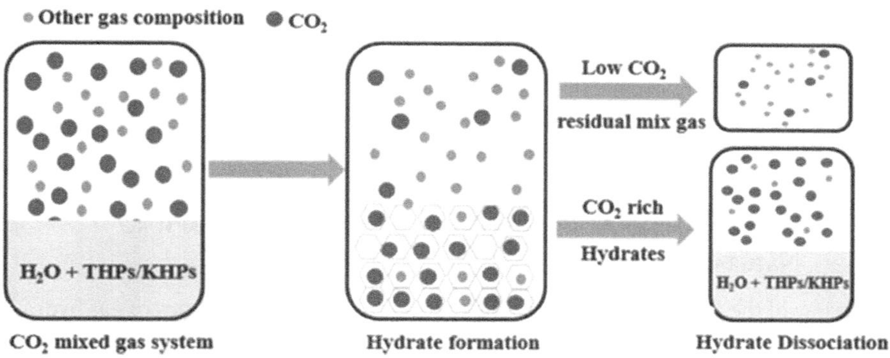

FIGURE 3.2 Hydrate-based gas separation process (CO_2 capture process).[30]

However, a review of ILs[25] indicates that many commonly used ILs for gas hydrate inhibition are toxic. In addition, ILs are relatively expensive and may not be cost-effective for use in the oil and gas industry.[26] In response to these challenges, amino acids were introduced as new gas hydrate inhibitors in 2011 by Sa et al.[37] They reported that amino acids exhibit strong electric charges and electrostatic interactions with water as zwitterions, interacting strongly with water molecules through hydrogen bonding due to their hydrophilic nature, which qualifies them as effective inhibitors. This electrostatic interaction between amino acids and water molecules disrupts the ice-like crystalline structure formed by hydrogen-bonded water molecules, thereby exerting an inhibitory effect.[28,29]

Generally, amino acids consist of carboxylic acid and amine groups, along with a side chain that can vary from apolar alkyl chains (hydrophobic) to moieties with positive or negative charges (hydrophilic), with their chemical and physical properties heavily influenced by these side chains.[31,32] Some key advantages of amino acids include their biologically friendly and biodegradable nature. Moreover, amino acids are cost-effective and can be purchased in bulk at relatively low prices. They have also been found to act as corrosion inhibitors for metals in various chemical systems, such as sulfuric acid and aqueous chloride solutions in nitric acid mediums, which alleviates corrosion concerns in field applications.[33,34] These properties make amino acids widely applicable in biological science, biotechnology, and the pharmaceutical industry, particularly for protein purification.[35,36] Importantly, these attributes also position amino acids as potential candidates for gas hydrate inhibition in pipelines. Notably, amino acids have been reported not only as gas hydrate inhibitors but also as effective gas hydrate promoters under both stirring and nonstirring conditions, making them promising candidates for future applications in CO_2 capture, gas separation, storage, and transportation.

The kinetics and thermodynamics of gas hydrates in the presence of amino acids play a crucial role in developing amino acid-based hydrate inhibitors and promoters. Research on gas hydrate-based applications involving amino acids is still in its early stages, with various studies focusing on their thermodynamic and kinetic effects. A critical review of existing data is therefore necessary, as there is currently no comprehensive review article in the open literature on the use of amino acids as gas hydrate promoters or inhibitors. This review aims to consolidate and analyze the reported findings from open literature on gas hydrate applications using amino acids. It will provide updated insights into the roles of amino acids as hydrate promoters and inhibitors, offering valuable information for future research and potential applications in hydrate-based technologies.

3.1.1 Role of Amino Acids in Hydrate Inhibition/CO_2 Capture/Natural Gas Storage

A review of the literature shows that the thermodynamics and kinetics of gas hydrates have been studied with amino acids. Most studies focus on the formation kinetics of gas hydrates, specifically in CO_2 capture/separation and gas storage. Researchers use the normal isochoric method with step heating for thermodynamic studies, while isothermal, constant cooling, and isochoric methods are used for kinetic studies. To analyze the data accurately, information on amino acids as gas hydrate additives was collected from available literature and examined separately for their thermodynamic and kinetic effects. Table 3.1 lists all gas hydrate systems studied with amino acids, including their tested concentrations and physicochemical properties (Tables 3.2–3.4).

3.2 ROLE OF AMINO ACIDS IN HYDRATE THERMODYNAMICS (PHASE BEHAVIOR)

3.2.1 Amino Acids as Thermodynamic Inhibitors

Generally, researchers determine the hydrate-liquid-vapor equilibrium (HLwVE) curve to evaluate the thermodynamic effect of amino acids as gas hydrate inhibitors or promoters. Seven amino acids (proline, glycine, alanine, arginine, serine, valine, and lysine) have been studied as THIs for CO_2, CH_4, and natural gas (NG: 93.0% CH_4, 5.0% C_2H_6, 2.0% C_3H_8).[27,54–58] Table 3.2 shows the experimental details of all reported HLwVE data measured with these amino acids.

Figures 3.3 and 3.4 show the HLwVE curve for CO_2, CH_4, and natural gas hydrates with amino acids at different concentrations in mol% and wt%. These figures demonstrate that adding amino acids shifts the HLwVE curve to higher pressure and lower temperature regions, indicating that all studied amino acids act as hydrate inhibitors in all gas systems. Notably, no thermodynamic hydrate promotion effect has been reported for amino acids in the literature. For CO_2 hydrates, the order of inhibition strength is valine > alanine > glycine, as shown in Figure 3.3(a). A similar trend is observed for CH_4 and natural gas systems in Figures 3.3(b) and 1(c). However, the inhibition strength based on mol% shows a decreasing order of proline, followed by serine, alanine, and glycine. Conversely, for CH_4 hydrates based on wt%, the inhibition strength is glycine > alanine > proline > serine > arginine, as shown in Figure 3.4. This variation in inhibition trends is due to the different concentration units used by various researchers. The concentration units used in gas hydrate studies are very important for evaluating and interpreting the impact of gas hydrate inhibition. Most reported studies on amino acids' thermodynamic effects on hydrates are

TABLE 3.1
List of various studied amino acids + studied gas systems, concentrations used and physicochemical properties

No.	Amino Acid	Gas	Side Chain Polarity	Side Chain	Hydropathy Index[d]	Type Test	Conc.	Remarks	Ref.
1	Glycine	CO_2	Nonpolar	–H	–0.4	THI	0.1[a] – 3.0[a]	Shows good thermodynamic hydrate inhibition impact	37
2	L-Alanine	CO_2	Nonpolar	–CH	1.8	THI	0.1[a] – 2.2[a]	Thermodynamically inhibit CO_2 hydrates	
3	L-Valine	CO_2	Nonpolar	$–CH(CH_3)_2$	4.2	THI	0.1[a] – 0.5[a]	Shows thermodynamic CO_2 hydrate inhibition	
4	Glycine	CO_2	Nonpolar	–H	–0.4	KHI	0.01[a] – 1.0[a]	Shows effective KHI impact by increasing the subcooling temperature and eliminates hydrates formation memory effect.	
5	L-Alanine	CO_2	Nonpolar	$–CH_3$	1.8	KHI	0.1[a]	Demonstrates kinetic hydrate inhibition impact but less efficient than glycine.	
6	L-Valine	CO_2	Nonpolar	$–CH(CH_3)_2$	4.2	KHI	0.1[a]	Shows less significant hydrate inhibition impact. Longer chains which are more hydrophobic do not inhibit hydrate. This is contrary to the understanding that hydrophobic compounds turn to be good KHIs (especially in ionic liquids (Tariq et al.,2014))	
7	Leucine	CO_2	Nonpolar	$–CH_2CH(CH_3)$	3.8	KHI	0.1[a]	Shows less significant hydrate inhibition impact	
8	Isoleucine	CO_2	Nonpolar	$–CH(CH_3)C_2H_5$	4.5	KHI	0.1[a]	Shows less significant hydrate inhibition impact	

(Continued)

TABLE 3.1
(Continued)

No.	Amino Acid	Gas	Side Chain Polarity	Side Chain	Hydropathy Index[d]	Type Test	Conc.	Remarks	Ref.
9	Glycine	CO_2	Nonpolar	–H	–0.4	Crystal structure	$0.1^a – 0.5^a$	Amino acids inclusion expands the hydrate crystal lattice, causing hydrate inhibition effect. At 2.2 mol% glycine's lattice expansion ability saturation is reached.	
10	L-Alanine	CO_2	Nonpolar	$–CH_3$	1.8	Crystal structure	$0.1^a – 0.5^a$	A structure I hydrate was formed with hydrate inhibition crystallization phenomenon. The lattice expansion magnitude was saturated at 0.5 mol%	
11	L-Valine	CO_2	Nonpolar	$–CH(CH_3)_2$	4.2	Crystal structure	$0.1^a – 0.5^a$	All amino acids have a distinct crystal structure. However, the inhibition strength of amino acids depends on whether they act individually or agglomerate during hydrate crystallization	
12	L-Alanine	CO_2	Nonpolar	$–CH_3$	1.8	KHI + spectroscopy	$0.01^a – 0.1^a$	Delays hydrate nucleation and growth rate via disruption of the water structure in hydrate formation.	
13	Aspartic acid	CO_2	Nonpolar	$–CH_2COOH$	–3.5	KHI + spectroscopy	$0.01^a – 0.1^a$	Delays hydrate nucleation and growth rate via disruption of the water structure in hydrate formation.	
14	Asparagine	CO_2	Nonpolar	$–CH_2CONH_2$	–3.5	KHI + spectroscopy	0.01^a	Delays hydrate nucleation and growth rate via disruption of the water structure in hydrate formation	

15	Phenylalanine	CO_2	Nonpolar	$-CH_2C_6H_5$	2.8	KHI + spectroscopy	0.1[a]	Relatively shows no effect on the nucleation kinetics of hydrate formation, especially in memory water, due to its water structure hydrogen bonding strengthening ability. However, it delays hydrate crystal growth which is incomparable to alanine	
16	Histidine	CO_2	Basic polar	$-CH_2C_3H_3N_2$	−3.2	KHI + spectroscopy	0.1[a]	Efficient in hydrate inhibition than alanine but less than aspartic acid and asparagine via disruption of the water structure in the hydrate formation	
17	Glycine	C_2H_6	nonpolar	−H	−0.4	KHI	0.05[b] – 3[b]		
18	Leucine	C_2H_6	nonpolar	$-CH_2CH(CH_3)$	3.8	KHI	0.05[b] – 3[b]		
19	Asparagine	CH_4	Polar	$-CH_2CONH_2$	−3.5	KHI + MD simulation			
20	Glycine	THF	nonpolar	−H	−0.4	KHI	0.05[b] – 1.5[b]		
21	Leucine	THF	nonpolar	$-CH_2CH(CH_3)$	3.8	KHI	0.05[b] – 1.5[b]	Inhibits hydrate formation kinetics but less than glycine.	
22	L-Threonine	CH_4	Polar	$-CH(OH)CH_3$	−0.7	KHI	2770[C] – 1385[C]	Shows no significant KHI effect in delaying hydrate nucleation in both fresh and memory system.	
23	L-Valine	CH_4	Nonpolar	$-CH(CH_3)_2$	4.2	KHI	2770[C] – 1385[C]	Shows no significant KHI effect in delaying hydrate nucleation in both fresh and memory system.	
24	L-Histidine	CH_4	Basic polar						
25	PVP and L-tyrosine	NG	Polar	−HO-Ph-CH_2	−1.3	KHI	1[b]	The presence of tyrosine improves the hydrate inhibition impact of NaCl + PVP system.	38

(Continued)

TABLE 3.1
(Continued)

No.	Amino Acid	Gas	Side Chain Polarity	Side Chain	Hydropathy Index[d]	Type Test	Conc.	Remarks	Ref.
26	PVP and L-Tyrosine	NG	Polar	–HO-Ph-CH_2	–1.3	KHI	100[c]–275[c]	Tyrosine is a strong inhibitor than PVP and its addition to PVP enhances hydrate nucleation time in several folds.	39
27	Glycine	CH_4	Nonpolar	–H	–0.4	THI	0.5[a]–3[a]	Inhibits methane hydrate phase boundary curve with increasing concentration.	40
28	Alanine	CH_4	Nonpolar	–CH_3	1.8	THI	0.5[a]–2.2[a]	Inhibits methane hydrate phase boundary curve with increasing concentration.	
29	Serine	CH_4	Polar	–HO–CH_2	–0.8	THI	1.3[a]–3[a]	Inhibits methane hydrate phase boundary curve with increasing concentration.	
30	Proline	CH_4	Nonpolar	–NH–$(CH_2)_3$	–1.6	THI	1.3[a]–9[a]	Inhibits methane hydrate phase boundary curve with increasing concentration.	
31	Glycine	CH_4	Nonpolar	–H	–0.4	KHI	0.1[a]	Exhibits hydrate nucleation time and growth rate delay in both fresh and memory water.	
32	Alanine	CH_4	Nonpolar	–CH_3	1.8	KHI	0.1[a]	Do not inhibit hydrate formation nucleation and growth rate.	
33	Serine	CH_4	Polar	–HO–CH_2	–0.8	THI	0.1[a]	Do not inhibit hydrate formation nucleation and growth rate.	
34	Proline	CH_4	Nonpolar	–NH–$(CH_2)_3$	–1.6	THI	0.1[a]	Do not inhibit hydrate formation nucleation and growth rate.	
35	Glycine	NG	Nonpolar	–H	–0.4	THI	0.5[a]–3[a]	Inhibits hydrate phase boundary curve with concentration.	

36	Alanine	NG	Nonpolar	$-CH_3$	1.8	THI	0.5[a]–2.2[a]	Inhibits hydrate phase boundary curve with concentration.	
37	Serine	NG	Polar	$-HO-CH_2$	−0.8	THI	1.3[a]–3[a]	Inhibits hydrate phase boundary curve with concentration.	
38	Proline	NG	Nonpolar	$-NH-(CH_2)_3$	−1.6	THI	1.3[a]–9[a]	Inhibits hydrate phase boundary curve with concentration.	
39	Glycine	NG	Nonpolar	–H	−0.4	KHI	0.1[a]	Exhibits hydrate nucleation time and growth rate inhibition effect.	
40	Alanine	NG	Nonpolar	$-CH_3$	1.8	KHI	0.1[a]	Do not inhibit hydrate formation nucleation and growth rate.	
41	Serine	NG	Polar	$-HO-CH_2$	−0.8	KHI	0.1[a]	Could inhibit hydrate formation kinetics better than glycine.	
42	Proline	NG	Nonpolar	$-NH-(CH_2)_3$	−1.6	KHI	0.1[a]	Do not inhibit hydrate formation nucleation and growth rate.	
43	Glycine	CO_2	Nonpolar	–H	−0.4	KHI	0.5[a]–3[a]	Inhibits hydrate formation rate with increasing concentration.	41
44	Proline	CO_2	Nonpolar	$-NH-(CH_2)_3$	−1.6	KHI	0.5[b]–2[b]	Inhibits hydrate formation rate with inhibition strength less than glycine but similar with serine and threonine.	
45	Serine	CO_2	Polar	$-HO-CH_2$	−0.8	KHI	0.5[b]–2[b]	Inhibits hydrate formation rate	
46	Threonine	CO_2	Polar	CH_3-CH(OH)	−0.7	KHI	0.5[b]–2[b]	Inhibits hydrate formation rate	
47	Glutamine	CO_2	Polar	$H_2NeCO-(CH_2)_2$	−3.5	KHI	0.5[b]–2[b]	Inhibits hydrate formation rate with the least inhibition strength compared with other studied amino acids.	
48	Histidine	CO_2	Basic polar	NH-CH= N-CH=C-CH_2	−3.2	KHI	0.5[b]–2[b]	Shows the highest hydrate formation inhibition impact compared with other studies amino acids.	

(Continued)

TABLE 3.1
(Continued)

No.	Amino Acid	Gas	Side Chain Polarity	Side Chain	Hydropathy Index[d]	Type Test	Conc.	Remarks	Ref.
49	Glycine	CH_4	Nonpolar	–H	–0.4	THI	5^b–20^b	Shows the highest hydrate phase behavior conditions inhibition compared with other studied amino acids.	42
50	Alanine	CH_4	Nonpolar	$–CH_3$	1.8	THI	10^b	Inhibits gas hydrate thermodynamically.	
51	Serine	CH_4	Polar	$–HO–CH_2$	–0.8	THI	10^b	Inhibits gas hydrate thermodynamically.	
52	Proline	CH_4	Nonpolar	$–NH–(CH_2)_3$	–1.6	THI	10^b	Inhibits gas hydrate thermodynamically.	
53	Arginine	CH_4	Basic polar	$HN{=}C(NH_2)$-NH-$(CH_2)_3$	–4.5	THI	10^b	Inhibits gas hydrate thermodynamically.	
54	Glycine	CO_2	Nonpolar	–H	–0.4	THI	5^b–20^b	Shows the highest hydrate phase behavior conditions inhibition compared with other studied amino acids.	43
55	Alanine	CO_2	Nonpolar	$–CH_3$	1.8	THI	10^b	Inhibits gas hydrate thermodynamically.	
56	Serine	CO_2	Polar	$–HO–CH_2$	–0.8	THI	10^b	Inhibits gas hydrate thermodynamically.	
57	Proline	CO_2	Nonpolar	$–NH$-$(CH_2)_3$	–1.6	THI	10^b	Inhibits gas hydrate thermodynamically.	
58	Arginine	CO_2	Basic polar	$HN{=}C(NH_2)$-NH-$(CH_2)_3$	–4.5	THI	10^b	Inhibits gas hydrate thermodynamically.	
59	L-Leucine	CH_4	Nonpolar	$–CH_2CH(CH_3)$	3.8	KHP/ morphology	0.1^b–0.5^b	Shows kinetic promotion with no promotion effect observed below 0.3wt%.	44
60	L-Methionine	CO_2	Nonpolar	CH_2-S-$(CH_2)_2^-$	1.9	KHP	0.02^b–1^b	Significantly promotes hydrate formation uptake without the use of energy-intensive mixing.	45

61	L-Norvaline	CO_2	Nonpolar	$C_{10}H_{19}NO_4$		KHP	0.02^b–1^b	Promotes hydrate formation with similar promotional impact as L-norleucine	
62	L-Norleucine	CO_2	Nonpolar	$C_6H_{13}NO_2$		KHP	0.02^b–1^b	Promotes hydrate formation	
63	2-Aminoheptanoic acid	CO_2	Acid	$C_7H_{15}NO_2$		KHP	0.02^b–1^b	Promotes hydrate formation but with less promotional impact compared with L-norleucine	
64	*n*-Hexanoic acid	CO_2	Acid	CH_3–4COOH		KHP	0.02^b–1^b	Promotes hydrate formation with similar promotional impact as 2-aminoheptanoic acid.	
65	n-hexylamine	CO_2	Nonpolar	$CH_3CH_2CH_2CH_2CH_2CH_2NH_2$		KHP	0.02^b–1^b	Promotes hydrate formation	
66	L-Tryptophan	CH_4	Nonpolar	Ph-NH-CH=C-CH_2^-	–0.9	KHP	0.01^b–0.3^b	Shows good kinetic hydrate formation enhancement effect in both stirred and unstirred systems.	46
67	L-Histidine	CH_4	Basic polar	NH-CH=N-CH=C-CH_2	–3.2	KHP	0.03^b–1^b	Shows hydrate formation promotional effect similar to arginine but less than tryptophan. Higher hydrophobic amino acids show less hydrate promotion effect.	
68	L-Arginine	CH_4	Basic polar	HN=C(NH_2)-NH-$(CH_2)_3$	–4.5	KHP	0.03^b–1^b	Shows hydrate formation promotion effect	
69	Lysine	CH_4	Basic polar	H_2N-$(CH_2)_4^-$	–3.9	THI	5^b–10^b	Shows THI effect with increasing concentration.	47
70	Lysine	CO_2	Basic polar	H_2N-$(CH_2)_4^-$	–3.9	THI	5^b–10^b	Shows THI effect with increasing concentration.	
71	Arginine	CH_4	Basic polar	HN=C(NH_2)-NH-$(CH_2)_3$	–4.5	THI/KHP	1^b–5^b	Slightly inhibits methane hydrate phase boundary as well as promoting hydrate formation uptake.	48
72	Valine	CH_4	Nonpolar	–CH$(CH_3)_2$	4.2	THI/KHP	1^b–5^b	Slightly inhibits methane hydrate phase boundary as well as promoting hydrate formation uptake.	

(Continued)

TABLE 3.1
(Continued)

No.	Amino Acid	Gas	Side Chain Polarity	Side Chain	Hydropathy Index[d]	Type Test	Conc.	Remarks	Ref.
73	Valine	CO_2	Nonpolar	$-CH(CH_3)_2$	4.2	THI/KHP	0.5[b]	Promotes hydrate formation uptake about 1.2 times.	49
74	Phenylalanine	CO_2	Nonpolar	$Ph\text{-}CH_2^-$	2.8	KHP	0.5[b]	Shows no significant hydrate promotional effect	
75	Cysteine	CO_2	Nonpolar	$HS\text{-}CH_2^-$	2.5	KHP	0.5[b]	Promotes hydrate formation uptake about 1.2 times.	
76	Methionine	CO_2	Nonpolar	$CH_3\text{-}S\text{-}(CH_2)_2-$	1.9	KHP	0.5[b]	Promotes hydrate formation uptake about 1.2 times.	
77	Threonine	CO_2	Polar	$CH_3\text{-}CH(OH)$	−0.7	KHP	0.5[b]	Shows no significant hydrate promotional effect	
78	Methionine	CO_2	Nonpolar	$CH_3\text{-}S\text{-}(CH_2)_2-$	1.9	KHP/XRD	0.5[b]	Shows efficient hydrate kinetics conversion rate, thus improving the hydrate formation uptake.	49
79	Phenylalanine	CO_2	Nonpolar	$Ph\text{-}CH_2^-$	2.8	KHP/XRD	0.5[b]	Shows less hydrate kinetics conversion rate, thus gives less hydrate formation uptake.	
80	Methionine	CH_4	Nonpolar	$CH_3\text{-}S\text{-}(CH_2)_2-$	1.9	KHP/XRD	0.5[b]	Shows efficient hydrate kinetics conversion rate, thus improving the hydrate formation uptake.	
81	Phenylalanine	CH_4	Nonpolar	$Ph\text{-}CH_2^-$	2.8	KHP/XRD	0.5[b]	Shows efficient hydrate kinetics conversion rate, hence enhances hydrate formation uptake.	

82	Methionine	CH_4 + CO_2	Nonpolar	CH_3-S-$(CH_2)_2$–	1.9	KHP/XRD	0.5[b]	Shows efficient hydrate kinetics conversion rate, hence enhances hydrate formation uptake.	
83	Phenylalanine	CH_4 + CO_2	Nonpolar	Ph-CH_2^-	2.8	KHP/XRD	0.5[b]	Shows efficient hydrate kinetics conversion rate, thus improving the hydrate formation uptake.	
84	Glycine + ethylene glycol	CH_4	Nonpolar	–H	–0.4	THI	1[b]–30[b] 1:1mixtures	Glycine can enhance the thermodynamic inhibition strength of ethylene glycol, indicating strong synergic inhibition effect.	
85	Glycine	CH_4	Nonpolar	–H	–0.4	MD simulation	0.45[b]–1.5[b]	Shows hydrate kinetics inhibition effect but less than serine.	
86	Alanine	CH_4	Nonpolar	–CH_3	1.8	MD simulation	0.45[b]–1.5[b]	Shows hydrate kinetics inhibition.	
87	Serine	CH_4	Polar	–HO–CH_2	–0.8	MD simulation	0.45[b]–1.5[b]	Shows efficient hydrate kinetics inhibition via interruption of the hydrogen bond network of water.	
88	Proline	CH_4	Nonpolar	–NH-$(CH_2)_3$	–1.6	MD simulation	0.45[b]–1.5[b]	Shows hydrate kinetics inhibition effect as alanine	
89	L-Leucine	CH_4 and NG	Nonpolar	–$CH_2CH(CH_3)$	3.8	KHP	0.1[b]–1[b]	Very efficient in promoting hydrate formation kinetics than all studied amino acids at low concentrations for both structure I and structure II natural gas hydrates systems.	50
90	L-Isoleucine	CH_4	Nonpolar	–$CH(CH_3)C_2H_5$	4.5	KHP	0.5[b]	Exhibits good hydrate promotional ability similar to phenylalanine.	
91	L-Valine	CH_4	Nonpolar	–$CH(CH_3)_2$	4.2	KHP	0.5[b]	Enhances hydrate formation kinetics.	
92	L-Threonine	CH_4	Polar	CH_3-CH(OH)	–0.7	KHP	0.5[b] –10[b]	Enhances hydrate formation with decreasing concentration.	

(Continued)

TABLE 3.1 (Continued)

No.	Amino Acid	Gas	Side Chain Polarity	Side Chain	Hydropathy Index[d]	Type Test	Conc.	Remarks	Ref.
93	L-Alanine	CH_4	Nonpolar	$-CH_3$	1.8	KHP	0.5^b-2^b	Exhibits negligible hydrate promotion effect with increasing concentration.	
94	L-Proline	CH_4	Nonpolar	$-NH-(CH_2)_3$	−1.6	KHP	0.5^b	Exhibits less hydrate promotion effect.	
95	L-Methionine	CH_4	Nonpolar	CH_3-S-$(CH_2)_2$-02.365	1.9	KHP	0.5^b	Shows good hydrate promotional impact.	
96	L-Tryptophan	CH_4	Nonpolar	Ph-NH-CH=C-CH_2^-	−0.9	KHP	0.5^b	Shows good hydrate promotional impact.	
97	L-Phenylalanine	CH_4	Nonpolar	Ph-CH_2^-	2.8	KHP	0.5^b	Shows good hydrate promotional impact.	
98	L-Arginine	CH_4	Basic polar	HN=C(NH_2)-NH-$(CH_2)_3$	−4.5	KHP	0.5^b	Able to promote hydrate formation kinetics with decreasing stability.	
99	L-Glutamic acid	CH_4	Basic polar	HOOC-$(CH_2)_2^-$	−3.5	KHP	0.5^b	Able to promote hydrate formation kinetics with decreasing stability.	
100	L-Histidine	CH_4	Basic polar	NH-CH=N-CH=C-CH_2	−3.2	KHP	0.5^b	Able to promote hydrate formation kinetics with decreasing stability.	
101	L-Serine	CH_4	Polar	−HO−CH_2	−0.8	KHP	0.5^b	Exhibits less hydrate promotional effect.	
102	L-Aspartic acid	CH_4	Acidic polar	$-CH_2COOH$	−3.5	KHP	0.5^b	Exhibits less hydrate promotional effect.	

[a] mol%; [b] wt.%; [c] ppm; [d] extracted from reference (Kyte and Doolittle, 1982)[70].

TABLE 3.2
Amino acids HL_wVE data

Author	Amino Acid	Gas	Conc./mol%	T/K	P/MPa	Data Points
		CO_2	0.1	274.55–281.35	1.49–3.51	5
		CO_2	0.5	274.35–281.05	1.49–3.50	5
		CO_2	1.3	273.85–280.65	1.49–3.51	5
		CO_2	2.2	273.35–280.15	1.44–3.48	5
	Glycine	CO_2	3	273.05–279.45	1.47–3.47	5
	Alanine	CO_2	0.1	274.55–281.45	1.49–3.52	5
		CO_2	0.5	274.25–280.95	1.48–3.49	5
		CO_2	1.3	273.75–280.35	1.47–3.49	5
		CO_2	2.2	273.25–279.95	1.46–3.48	5
	Valine	CO_2	0.1	274.45–281.35	1.48–3.51	5
37		CO_2	0.5	274.15–280.85	1.48–3.50	5
40	Glycine	CH_4	0.5	274.45–284.85	2.940–8.965	5
		CH_4	1.3	273.95–284.30	2.953–8.93	5
		CH_4	2.2	273.35–283.75	2.942–8.923	5
		CH_4	3	272.85–283.05	2.916–8.871	5
		NG	0.5	276.25–286.75	1.248–4.086	5
		NG	1.3	275.85–286.45	1.243–4.103	5
		NG	2.2	275.45–285.95	1.247–4.088	5
		NG	3	274.85–285.35	1.245–4.07	5
	Alanine	CH_4	0.5	274.25–284.85	2.947–8.952	5
		CH_4	1.3	273.95–284.15	2.953–8.928	5
		CH_4	2.2	273.05–283.58	2.932–8.914	5
		NG	0.5	276.15–286.65	1.251–4.102	5
		NG	1.3	275.75–286.35	1.245–4.106	5
		NG	2.2	285.75–275.15	1.237–4.086	5
	Serine	CH_4	1.3	273.75–284.05	2.938–8.94	5
		CH_4	3	272.65–282.85	2.937–8.889	5
		NG	1.3	274.85–285.45	1.241–4.066	5
		NG	3	273.65–283.75	1.234–4.055	5
	Proline	CH_4	1.3	283.85–273.65	8.934–2.941	5
		CH_4	3	272.3–282.50	2.929–8.868	5
		CH_4	6	268.40–278.65	28.87–8.698	5
		CH_4	9	264.90–274.00	2.839–8.473	5
		NG	1.3	274.85–285.45	1.241–4.066	5
		NG	3	273.65–283.75	1.234–4.055	5
		NG	6	270.75–280.65	1.235–3.995	5
		NG	9	267.65–276.75	1.206–3.932	5
51	Glycine	CH_4	5 wt%	277.90–285.20	4.550–9.840	4
		CH_4	10 wt%	277.25–284.50	4.650–9.980	4
		CH_4	15 wt%	276.80–283.73	4.600–9.650	4
		CH_4	20 wt%	276.50–283.10	4.800–9.770	4

(Continued)

TABLE 3.2
(Continued)

Author	Amino Acid	Gas	Conc./mol%	T/K	P/MPa	Data Points
	Alanine	CH_4	10 wt%	277.55–284.30	4.605–9.550	4
	Serine	CH_4	10 wt%	277.70–285.00	4.595–9.800	4
	Proline	CH_4	10 wt%	277.60–284.85	4.550–9.820	4
	Arginine	CH_4	10 wt%	278.55–285.40	4.700–9.650	4
43	Glycine	CO_2	5 wt%	278.30–281.45	2.600–3.980	4
		CO_2	10 wt%	277.60–280.70	2.610–3.960	4
		CO_2	15 wt%	276.60–279.80	2.550–3.960	4
		CO_2	20 wt%	275.60–279.20	2.520–3.960	4
	Alanine	CO_2	10 wt%	277.60–280.87	2.560–4.000	4
	Serine	CO_2	10 wt%	278.20–281.30	2.600–4.000	4
	Proline	CO_2	10 wt%	277.70–281.10	2.530–4.020	4
	Arginine	CO_2	10 wt%	278.30–281.50	2.560–3.970	4
47	Lysine	CO_2	5 wt%	276.20–281.80	2.200–4.010	4
		CH_4	10 wt%	276.45–281.03	2.000–4.010	4
		CH_4	5 wt%	278.15–285.62	4.600–10.01	4
		CH_4	10 wt%	278.05–285.20	4.900–10.40	4
48	Arginine	CH_4	5 wt%	278.80–285.90	4.550–9.840	4
	Valine	CH_4	5 wt%	278.60–285.80	4.600–9.650	4
52	Glycine + ethylene glycol	CH_4	0.5 wt% + 0.5 wt%	279.70–287.80	5.050–12.20	5
		CH_4	2.5 wt% + 2.5 wt%	279.10–286.70	5.110–11.98	5
		CH_4	5 wt% + 5 wt%	277.10–285.40	4.780–11.47	5
		CH_4	10 wt% + 10 wt%	274.70–282.20	4.880–11.47	5
		CH_4	15 wt% + 15 wt%	273.30–279.90	4.810–11.15	5
53	Valine	CH_4	1 wt%	276.20–284.10	3.600–8.10	4
			5 wt%	275.70–283.50	3.500–8.00	4
	Threonine	CH_4	1 wt%	278.60–286.00	4.600–10.10	4
			5 wt%	277.00–285.70	4.000–10.20	4
	Asparagine	CH_4	1 wt%	277.90–286.10	4.300–10.30	4
			5 wt%	275.80–283.70	3.500–8.10	4
	Phenylalanine	CH_4	1 wt%	276.20–284.00	3.600–8.20	4
			5 wt%	275.90–283.90	3.600–8.00	4
	Glycine + 1-ethyl-3-methy-limidazoliumchloride	CH_4	5 wt% + 5 wt%	277.80–284.90	4.700–9.99	4

TABLE 3.3
Variations in some studied amino acids concentration units

Wt%	Mol%				
	Glycine	Alanine	Proline	Serine	Valine
5	1.25	1.05	0.82	0.89	0.80
10	2.60	2.20	1.71	1.87	1.68
15	4.06	3.45	2.69	2.94	2.64
20	5.66	4.81	3.76	4.11	3.70

TABLE 3.4
Comparison of the KHI/KHP efficiency of amino acids and conventional additives

Amino Acid	Remarks	Reference
	Commercial KPIs (SDS)	
Histidine	SDS promotes methane hydrate better than histidine at 1 wt%.	71
Leucine	Leucine is not efficient as SDS in promoting methane hydrate at 0.3 wt%.	86
Valine	Valine is an effective methane hydrate promoter than SDS at 1 wt%.	20
Arginine	Arginine is a poor promoter of methane hydrate compared with SDS at 1 wt%.	20
Histidine	SDS is a good promoter than histidine for ethane hydrate formation. However, histidine effectively promoters methane + propane hydrate than SDS.	105
	Commercial KHIs (PVP/PVCap)	
Glycine	Glycine and PVP has similar CO_2 hydrate inhibition impact efficiency.	78
Tyrosine	PVP is efficient than tyrosine in preventing natural gas hydrate at 1 wt%.	72
Tyrosine	PVP is a poor inhibitor compared to tyrosine for methane + ethane hydrate at 0.02 wt%.	75
Histidine	Histidine is more efficient than PVP in preventing CO_2 hydrate formation at 1.5 wt%, but similar at 1 wt%.	41
Glycine	PVP is slight better than glycine.	41
Glycine	Glycine exhibits weak hydrate formation inhibition impact compared to PVP (for pure ethane and mixed methane + propane).	105
Glycine	PVCap is more efficient in prevention CH_4 hydrate formation than glycine at 1 wt%.	79

measured in mol%.[27,58] Figures 3.3 and 3.4 show that the equivalent concentrations in mol% and wt% of amino acids reveal significant differences in inhibition trends. This difference can affect the analysis of their inhibition impact depending on which concentration unit is used. As suggested by Mech et al.,[60] using different units could lead to opposing observations or reports on inhibition impact. For instance, when mol% is used, amino acids with heavier molecular weights (longer side chains) tend to

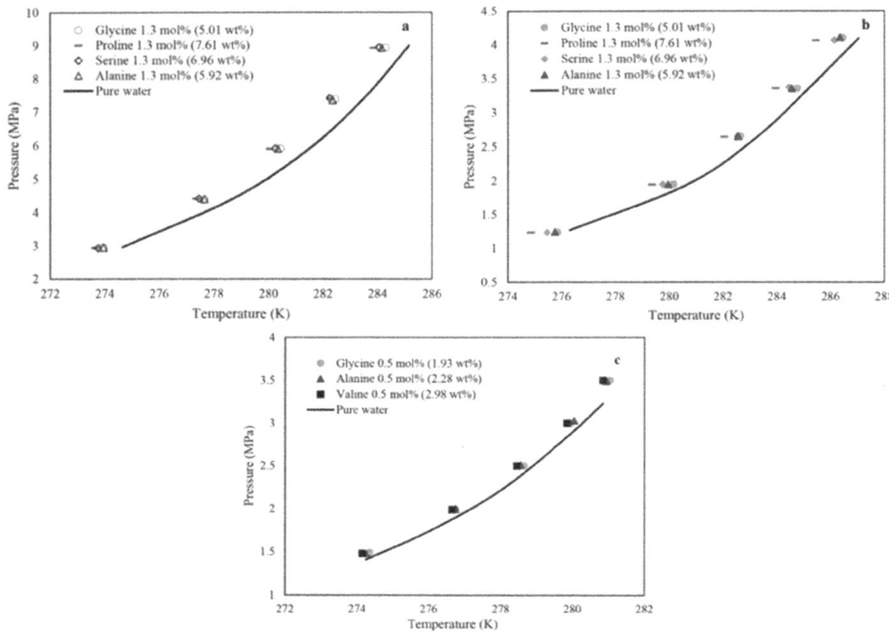

FIGURE 3.3 The inhibition strength of amino acids on the HLwVE curve in various gas systems showing the effect of concentration units studied on inhibition impact.[59]

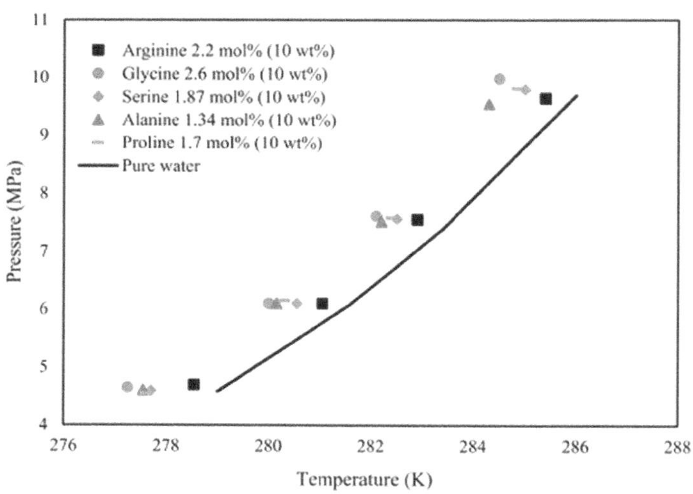

FIGURE 3.4 The inhibition impact of amino acids on the HLwVE curve of CH_4 hydrate systems showing the effect of studied concentration units on inhibition impact.[62]

exhibit higher inhibition, and vice versa. This can be clearly seen in Table 3.3, where the equivalent wt% concentrations of amino acids in mol% are relatively low, especially for amino acids with higher molecular weights. Conversely, based on wt%, the hydrate inhibition effect increases as the molecular weight decreases (shorter side chain length), as depicted in Figures 3.3 and 3.4. In many industrial applications, wt% is the preferred unit of measurement.[61] Therefore, for research focused on industrial applications, using wt% might be more appropriate. This choice of unit allows for interpretations that are more directly applicable to practical field scenarios.

Based on wt%, glycine is the most effective amino acid THI. Long et al.[63] found that glycine can also enhance the thermodynamic inhibition performance of ethylene glycol (a commercial THI) on CH_4 hydrates. They reported that a 20 wt% glycine solution shows a methane hydrate phase boundary deviation temperature of 2.9 K,[55] while a combination of 10 wt% glycine and 10 wt% ethylene glycol shows a deviation of 5.2 K,[63] as shown in Figure 3.5. Interestingly, the inhibition effect of 5 wt% glycine plus 5 wt% ethylene glycol is similar to that of 10 wt% glycine. Thus, glycine's enhancement of ethylene glycol's inhibition is more evident at concentrations above 5 wt%. However, a mixture of glycine and 1-ethyl-3-methylimidazolium chloride (an IL) at 10 wt% (50/50) has negligible effect on their phase behavior compared to their pure compositions at the same concentration.[56]

In addition, lysine shows similar inhibition to alanine for methane and carbon dioxide hydrates at 10 wt%.[57] Valine shows less inhibition for methane and carbon dioxide hydrates, likely due to its longer alkyl side chain.[27] The thermodynamic

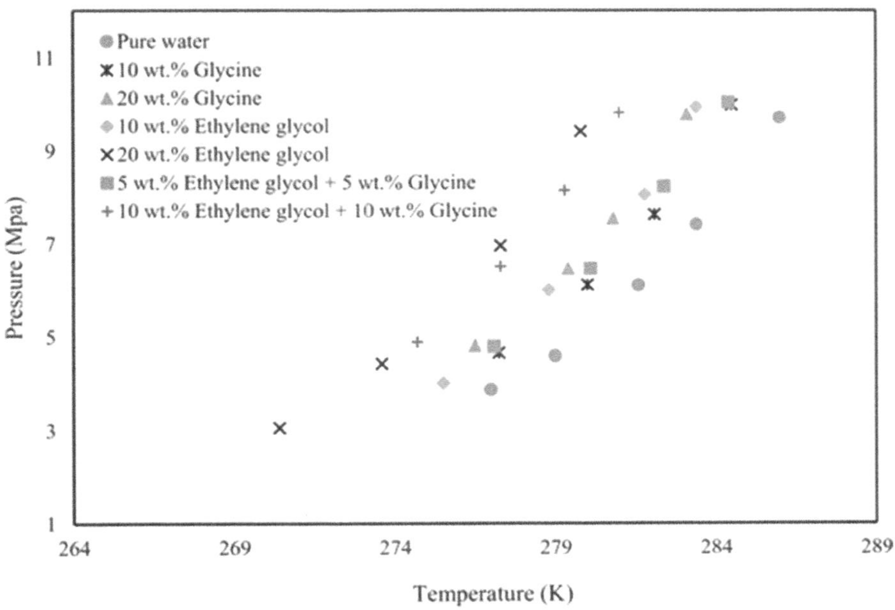

FIGURE 3.5 The inhibition impact of pure glycine and glycine + ethylene glycol on the HLwVE data of CH_4 hydrates.[59]

effects of threonine, valine, phenylalanine, and asparagine are not as strong as those of glycine and alanine at 5 wt% for CH_4 hydrate formation.[23] However, these amino acids are mostly methane hydrate kinetic promoters. For instance, in carbon dioxide hydrate systems, asparagine and phenylalanine act as promoters, with phenylalanine also promoting CH_4 hydrate.[64,65] Similarly, threonine and valine promote CH_4 hydrates kinetically.[20,64,66]

The thermodynamic inhibition mechanism of amino acids is due to their electrostatic interactions via zwitterion interactions and hydrogen bonding with water molecules, which disrupts water's role in hydrate formation and results in inhibition.[27,55,65] An ANOVA analysis at the 95% confidence level indicated that the thermodynamic inhibition impact of amino acids is not dependent on the type of guest compound (for methane and carbon dioxide systems) and is solely due to their molecular interactions with water in the liquid phase. The inhibition strength of amino acids on gas hydrate phase behavior is influenced by their hydrophobicity, solubility in water, side chain length, and concentration.[27] All tested amino acids show increased inhibition with increasing concentration.[27,55]

The affinity of each natural amino acid for water has been evaluated based on various physicochemical and interaction properties. These studies have led to the development of amino acids' side chain hydrophobic scales. There are several such scales available[67] because authors study different amino acid properties (e.g., surface tension, solubility, accessible surface areas, energy of transfer of amino acids from water to a less polar environment) to determine their hydrophobicity. Some authors[27,65,68] have suggested that the inhibition effect of amino acids on gas hydrates is influenced by their hydropathy/hydrophobicity. The hydropathy of compounds significantly affects their gas hydrate inhibition strength. This relationship is well established in ILs, where hydrate inhibition increases with decreasing hydropathy, which is related to the alkyl chain length of the compounds.[69]

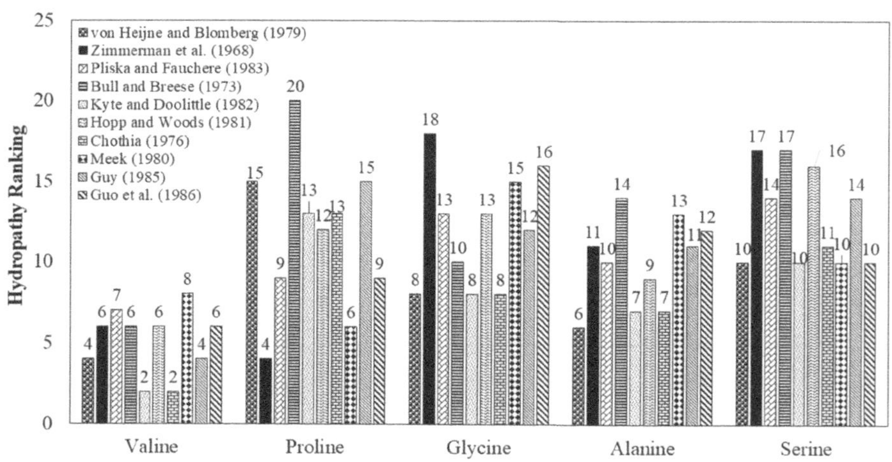

FIGURE 3.6 Hydropathy ranking of studied gas hydrate inhibition. Data are taken from Wilce et al.[106]. The hydropathy of amino acids decreases with increasing ranking number.

Notwithstanding, there are several amino acid hydropathy scales available in the literature, as summarized in Figure 3.6. However, there is little agreement among these scales, indicating that amino acid hydropathy is not well understood. This inconsistency makes it difficult to select a suitable hydropathy scale for gas hydrate data analysis, which can lead to misinterpretation of results or errors in gas hydrate data analysis.

The hydropathy of an amino acid refers to its hydrophilicity and hydrophobicity. This describes the ability of amino acids to interact with water molecules or prevent their interaction with water.[70] Studying amino acids' hydropathy is challenging because various scales exist based on different properties such as solubility and surface tension. These scales assign numbers to each amino acid to describe its hydropathy strength. Higher hydropathy values indicate strong hydrophobicity, while lower values indicate strong hydrophilicity.

Generally, gas hydrate researchers[27,65] use the amino acid hydropathy scales developed by Kyte and Doolittle.[70] The reasons for choosing these scales are not mentioned, but it might be because they are the most commonly used hydropathy scales in literature. Figure 3.7 illustrates the correlation between the effectiveness of amino acids in inhibiting gas hydrates (measured by average temperature depression) and their hydropathy scale proposed by Kyte and Doolittle.[70] In Figure 3.7(a), the R^2 values are 0.46 and 0.38 for methane and natural gas hydrate inhibition, respectively, while in Figure 3.7(b), the R^2 value is 0.67 for methane. This indicates that the hydrate inhibition strength of amino acids does not have a strong correlation with their hydropathy as shown in Figure 3.7. Meanwhile, researchers generally use this hydropathy scale as the basis for analyzing the impact of amino acids on hydrate inhibition.[27] However, this approach can be misleading and might lead to data analysis errors. Therefore, we recommend further studies to select or develop a more suitable amino acid hydropathy scale for hydrate inhibition purposes. It should be noted that the R^2 values in Figure 3.7 might be influenced by the number of data points used in the correlation analysis, as there is limited data available in the literature. Consequently, more experimental hydrate phase equilibrium data of amino acids are needed to fully understand the effect of amino acid hydropathy on their

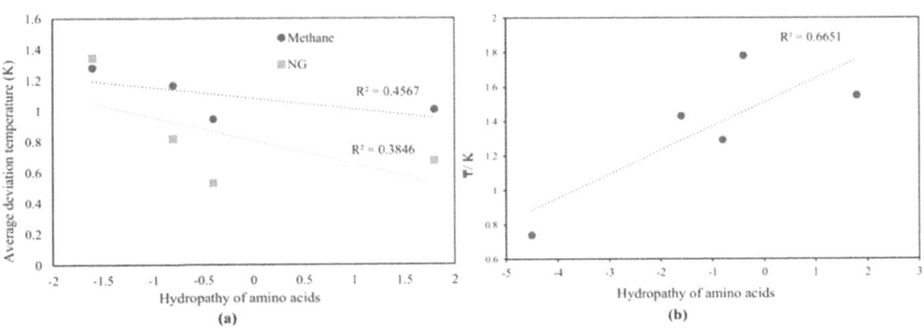

FIGURE 3.7 Regression between average depression temperature (Ŧ) and commonly used amino acid hydropathy scale proposed by Kyte and Doolittle (1982); (a) data from Sa et al. (2016) and (b) data from Bavoh et al. (2016b).

inhibition impact. For instance, serine is less effective in preventing hydrate formation than glycine, despite having a lower hydropathy value (−0.8 for serine compared to 0.4 for glycine). Thus, relying solely on the hydropathy scale to explain the hydrate inhibition effect of amino acids is insufficient. Other characteristics, such as amino acids pH levels (acidity), side chain polarity, and side chain group type (e.g., acyclic, aliphatic, aromatic, containing sulfur or hydroxyl), should be critically considered when discussing the inhibition or promotion impact of amino acids on gas hydrate formation.

The solubility of THIs in water is crucial for inhibiting gas hydrate formation. Conditions such as low temperatures during hydrate formation and acidic environments in solutions caused by dissolved gases like carbon dioxide decrease the solubility of amino acids. Sa et al.[27] used the van Hoff equation to determine the solubility of amino acids and account for the solubility reduction due to the acidic environment. They suggested that the solubility reduction of amino acids due to an acidic environment is negligible and that only the effect of decreasing temperature should be considered. Therefore, the hydrate inhibitory efficiency of each amino acid increases with concentration within their respective solubility limits in water.

3.3 ROLE OF AMINO ACIDS IN HYDRATE KINETICS

3.3.1 Amino Acids as Kinetic Inhibitors

Unlike thermodynamic studies, there are relatively many studies on the kinetics of amino acids in gas hydrate mitigation or enhancement. Kinetic data are treated differently because gas hydrate formation kinetics are highly probabilistic and depend on factors such as apparatus design, experimental procedure, reactor wall roughness, driving force, and impurities in the sample. Generally, the three main kinetic indicators used to evaluate the inhibition or promotion performance of amino acids are nucleation time, rate of hydrate formation, and gas uptake during hydrate formation. For evaluating hydrate inhibition, nucleation time is preferred as it indicates the efficiency of amino acids in delaying hydrate formation. In contrast, for hydrate promotion applications, the rate of hydrate formation and total gas uptake are primarily considered. Based on kinetic measurements and studies, we observed that amino acids are generally poor kinetic inhibitors of gas hydrates; they act more as kinetic promoters. However, their inhibition strength lies in their ability to delay the growth rate of hydrate formation and gas uptake. The kinetic inhibition parameters typically determined by researchers include induction time[71–75] and onset hydrate formation temperature (subcooling temperature).[58,72,76] In addition, some authors have determined gas uptake[58,65,71,72,77,78] and the rate of hydrate formation.[77] Sa et al.[78] studied the effect of five amino acids (alanine, glycine, leucine, valine, and isoleucine) on CO_2 hydrates at 0.1 mol% by measuring their subcooling temperature and gas uptake for fresh and memory water systems. Their findings showed that glycine was the most effective at inhibiting CO_2 hydrates, followed by alanine, valine, leucine, and isoleucine. Moreover, the inhibition effect of glycine increased with higher concentrations. Sa et al.[65] extended their study on the inhibition impact of amino acids on CO_2 hydrate formation growth and nucleation kinetics at 0.01 and 0.1 mol% using five amino acids with electrically charged and/or hydrophilic side chains: alanine, asparagine, aspartic acid, histidine, and phenylalanine. At 0.01 mol%, asparagine and aspartic acid were more efficient inhibitors than alanine based on gas uptake, while at 0.1 mol%, histidine showed the strongest inhibition effect, followed by alanine and phenylalanine. According to Sa et al.,[65] the trends in hydrate nucleation and growth inhibition of these amino acids correlate with their hydropathy index at both low (0.01 mol%) and high (0.1 mol%) concentrations. In addition, histidine performed better than alanine in delaying CO_2 hydrate nucleation time and growth. However, phenylalanine was less efficient in preventing hydrate formation compared to alanine, and had no significant impact on delaying the hydrate nucleation process. Interestingly, unlike glycine in Sa et al.'s[78] previous study, the inhibition impact of aspartic acid and asparagine decreased with increasing concentration due to their solubility limitations, leading to residual excess (unreacted) amino acid in the system, which serves as sites for enhancing hydrate formation, thus reducing their kinetic inhibitory efficiency. Roosta et al.[77] reported that the kinetic inhibition effect of amino acids on CO_2 hydrates is due to their side chain hydrophobicity and electrical charge. Therefore, histidine showed a higher inhibition impact than glycine, followed by proline, whose inhibition strength is in the same range as serine and threonine but higher than glutamine. It must be stated that the correlation between the amino acids' side chain properties and inhibition impact is not well understood and requires further studies. However, amino acids with polar side chains generally seem to show better CO_2 hydrate inhibition than nonpolar ones.

Perfeldt et al.[76] reported that valine exhibits slightly higher CH_4 hydrate inhibition than threonine and is more effective than some antifreeze proteins. However, a recent study showed that glycine, serine, proline, and alanine could inhibit methane and natural gas (93% CH_4, 5% C_2H_6, 2% C_3H_8) hydrates at 0.1 mol%, based on onset temperature and gas uptake evaluation. Among these, proline was the most effective. Talaghat[75] suggested that tyrosine could delay the induction time of natural gas hydrate better than PVP in a mini flow loop apparatus at 200 ppm. Furthermore, they argued that adding tyrosine to PVP increased the inhibition impact of PVP. A study by Kakati et al.[72] reported that the incorporation of tyrosine synergistically with PVP boosts the kinetic inhibition efficiency of PVP for natural gas hydrate systems. Xu et al.[79] argued that glycine, when evaluated through methane hydrate formation kinetics, performs poorly in mitigating hydrate formation compared to PVCap. However, it can significantly improve the efficiency of PVCap, increasing it by about 16 times. This demonstrates the ability of amino acids to inhibit gas hydrates while also enhancing the performance of conventional kinetic inhibitors in the oil and gas industry. Despite their poor performance in delaying hydrate nucleation time when used alone, amino acids can increase the induction time of conventional kinetic inhibitors when mixed together. In the presence of THF and C_2H_6 hydrates, glycine is considered a stronger KHI than L-leucine.[73] Thus, glycine stands out among the studied amino acids as the best kinetic inhibitor in various hydrate-forming systems.

Conversely, amino acids have been utilized as promoters for gas hydrate dissociation (inhibition) in methane hydrate production. Kumar et al.[108] have patented a method for natural methane hydrate recovery using amino acids, including glycine, histidine, proline, tyrosine, serine, threonine, and tryptophan. The patent asserts that all tested amino acids effectively enhance methane hydrate dissociation kinetics after 18 minutes at 283 K compared to a control sample (pure water). Notably, in a stirred reactor, glycine and histidine exhibit a significant impact on hydrate dissociation enhancement. Histidine typically demonstrates high methane recovery after 30 min, whereas proline is the least effective in promoting methane hydrate dissociation. However, histidine's efficiency does not surpass that of ethylene glycol, a commercial hydrate thermodynamic inhibitor, as ethylene glycol more effectively destabilizes the hydrate phase. Furthermore, methane recovery improves with an increased injection rate of the additives (amino acids) at 10 and 30 ml/min.

3.3.2 Amino Acid Kinetic Inhibition Mechanism

It is generally accepted that commercially utilized gas hydrate kinetic inhibitors, such as polymers, function through adsorption. However, Sa et al.[78] proposed a different inhibition mechanism for amino acids by examining their effect on CO_2 hydrate using synchrotron powder X-ray diffraction (PXRD) to determine the crystal structure of CO_2 hydrates and their lattice parameters. It was hypothesized that amino acids may inhibit hydrate growth through a mechanism distinct from that of PVP, which is primarily driven by adsorption. This alternative mechanism involves the perturbation of the local water structure by the hydrophilic terminal groups and hydrophobic side chains of amino acids via hydrogen bonding, as illustrated in Figure 3.8(a). Further research by Sa et al.[65] investigated the perturbation effect of amino acids on the local water structure by analyzing the polarized Raman spectra of aqueous amino acid solutions. Their findings indicated that amino acids disrupt the structure of liquid water, leading to the kinetic inhibition of gas hydrate nucleation and growth. The extent of this perturbation depends on the properties of the amino acid side chains; amino acids with electrically charged and/or hydrophilic side chains were observed to disrupt the low-temperature liquid water structure, whereas those with hydrophobic side chains reinforced it. In addition, Sa et al.[80] studied the crystallization phenomena of CO_2 hydrate in the presence of amino acids using PXRD, 13C cross-polarization (CP) nuclear magnetic resonance (NMR), and Raman spectroscopy. Their results contradicted the previously proposed gas hydrate mitigation mechanism involving the perturbation of the local water structure.[65,78] It was found that amino acids form hydrogen bonds with water molecules, thereby displacing water molecules within the hydrate crystal lattice and incorporating themselves into the hydrate structure. This incorporation leads to lattice distortion and expansion. However, once the lattice sites for incorporation are saturated, the amino acids that are not incorporated into the hydrate crystal lattice are excluded and crystallize among themselves. These excluded crystallized amino acids may act as sites that enhance gas hydrate formation. It is important to note that amino acids do not form semi-clathrate hydrates; they only participate in lattice formation (see Figure 3.8(b)). This has been confirmed by estimating the hydrate enthalpy of dissociation using the Clausius-Clapeyron equation, which indicates that amino acids do not participate in hydrate cage occupation and structure during hydrate formation.[54,55] It must be noted that the findings of references[65,78,80] require more direct evidence and further molecular-level confirmations to elucidate the mechanism of amino acid hydrate

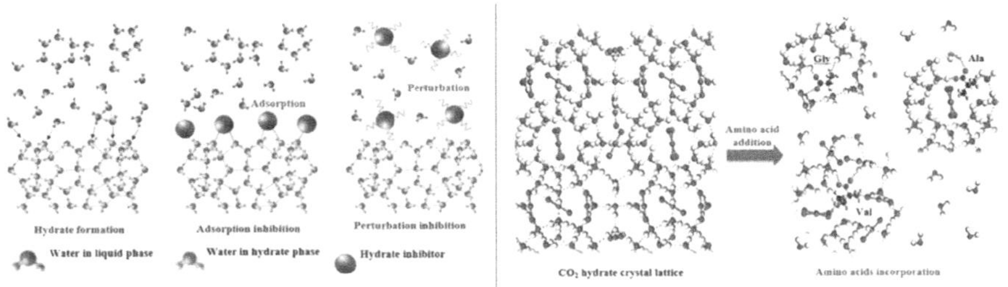

FIGURE 3.8 (a) Amino acids gas hydrate growth inhibition mechanism by perturbation of the local water structure compared to adsorption inhibition mechanism (Sa et al., 2013); (b) amino acids lattice distortion and expansion inhibition mechanism through incorporation into gas hydrate crystal lattice (Sa et al., 2014). ©Nature Publishing Group. Reproduced by permission of Nature Publishing Group.

inhibition. Their studies primarily relate the ice lattice Bragg peaks to structure I (sI) hydrates, which may reflect the water-to-hydrate conversion rate in the system, potentially influenced by factors such as system driving force (particularly at 3.6 MPa for CO_2 hydrates), stirring rate, gas-to-water ratio, and reactor design. Moreover, Sa et al.'s[80] study on lattice incorporation lacks quantitative analyses and provides limited crystalline information, focusing only on profile refinement. Therefore, a careful analysis of the lattice incorporation phenomena of amino acids in the hydrate lattice structure is necessary, as this incorporation could adversely affect or change various lattice refinement parameters, including lattice parameters (a, b, c, α, β, γ), atomic site occupancies, atomic positions (x, y, z), and profile parameters (U, V, W), potentially altering the structure.

In addition, the notion of amino acids incorporating into hydrate lattice structures is expected to result in a thermodynamic inhibition effect rather than the kinetic inhibition suggested by Sa et al.[80] This discrepancy might be attributed to the previously discussed perturbation kinetic inhibition mechanism. Essentially, both the thermodynamic inhibition effect and the perturbation kinetic hydrate inhibition mechanism are driven by hydrogen bonding interactions between the hydrogen-bonded water crystalline structure and the amino acid molecules. Consequently, a significant perturbation effect occurs, which kinetically reduces the hydrate crystalline nucleation and growth rate.

3.3.3 Amino Acids as Kinetic Promoters

Gas hydrate promoters are additives that enhance hydrate formation either thermodynamically or kinetically. These additives are crucial for the implementation of gas hydrate-based technologies, such as natural gas storage and transportation, as well as CO_2 capture, storage, and sequestration. A critical challenge that limits the implementation of these technologies is achieving rapid hydrate formation. Conventional gas hydrate promoters include THF[8,81,82] and sodium dodecyl sulfate (SDS).[83] However, these promoters do not facilitate hydrate formation at the speed required for practical applications. In addition, they are not environmentally friendly and can lead to foam formation in process plants.[84] Recent studies suggest that amino acids are potential gas hydrate promoters. Most importantly, amino acids do not promote foam formation, making them suitable for application in commercial hydrate-based operations.[84]

This section reports only on kinetic amino acid-based hydrate promoters. Liu et al.[85] were among the first to report natural amino acids as methane hydrate promoters at low concentrations up to 1 wt%. Their study found that leucine demonstrated the highest CH_4 hydrate promotion effect compared to methionine, tryptophan, phenylalanine, arginine, glutamic acid, and histidine at 0.5 wt%. Leucine was able to convert approximately 95% of water into methane hydrate with a gravimetric capacity of 144 mg/g at an optimal concentration of 0.5 wt%. Importantly, leucine did not cause foaming upon degassing. However, L-serine, L-aspartic acid, L-proline, and alanine showed very low methane hydrate uptake, behaving as inhibitors as demonstrated by Sa et al.[58] Further details on the morphology changes of leucine during methane hydrate formation and dissociation were studied by Veluswamy et al.,[86] although no hydrate enhancement effect was detected below 0.3 wt%.

Veluswamy et al.[84] further demonstrated that tryptophan could promote methane hydrate formation more effectively than histidine and arginine, but not as effectively as leucine. They argued that the properties of amino acid side chains play a critical role in hydrate promotion, with amino acids containing aromatic side chains enhancing hydrate formation better than those with aliphatic side chains. The combination of aromatic and hydrophobic side chains could better promote hydrate formation. This effect is particularly relevant for methane hydrates, as the promotion effect of amino acids is composition-dependent. All studied amino acids with aromatic side chains and hydrophobic nature (tryptophan, leucine, phenylalanine) showed significant methane hydrate promotion. However, leucine exhibited a poor promotion effect (inhibition effect) in ethane and THF hydrates.[73,74]

Similarly, phenylalanine has been reported to slightly inhibit CO_2 hydrate formation kinetics.[65] In addition, histidine has been shown to exhibit a kinetic promotion effect on CH_4 hydrate.[71] Conversely, histidine has also been reported to kinetically inhibit CO_2 hydrates,[65,77] indicating that the kinetic promotion or inhibition effect of amino acids significantly depends on the type of guest compound present. This composition-dependent hydrate promotion effect of amino acids provides opportunities for selectivity in gas hydrate-based mixed gas separation and CO_2 capture applications. Interestingly, tryptophan and methionine are able to promote the formation of both CH_4 and CO_2 hydrates.[87] Other factors contributing to the promotion or inhibition effect of amino acids include their side chain length and hydropathy index.

Authors assert that there is an optimal side chain length for hydrophobic amino acids in the context of hydrate kinetic promotion or inhibition.[78,87] However, the precise optimal side chain length is not clearly defined in current studies. According to Cai et al.,[87] L-methionine could promote CO_2 hydrate formation more effectively than L-norvaline, L-norleucine, 2-aminoheptanoic acid, *n*-hexanoic acid, and *n*-hexylamine at a concentration of 0.2 wt%. The gravimetric capacity of CO_2 hydrate formation was approximately 356 mg/g in 1000 min for an 81 mg/g bulk water system. It is noteworthy that the promotion effect of amino acids is concentration-dependent, varying for each amino acid in different gas systems. For each gas system, all amino acids have an optimal concentration above which their promotion or inhibition impact decreases. For instance, the optimal promotion impact of leucine in CH_4 hydrate is in the range of 0.3–0.5 wt%.[85,86] In the CH_4 hydrate system, the optimal concentration for tryptophan is 0.3 wt%, while for histidine and arginine it is 1 wt%.[84] In the CO_2 hydrate system, L-methionine has an optimal concentration of 0.2 wt%.[87] It is recommended that authors optimize the effective promotion or inhibition concentration for amino acids and compare them accordingly.

In Bhajan's lab, the effect of valine and arginine on CH_4 hydrates was studied, showing that both amino acids promote CH_4 hydrate formation more effectively than SDS. Valine exhibited the most efficient methane hydrate promotion, with CH_4 consumption rates approximately ten times higher than pure water and 1.3 times higher than SDS. However, the induction time for CH_4 hydrate nucleation was shorter compared to SDS.

Prasad and Kiran[66] investigated the effect of five amino acids (L-valine, L-phenylalanine, L-cysteine, L-methionine, and L-threonine) on CO_2 hydrate formation under isochoric conditions in both stirring and nonstirring modes. They found that L-valine, L-cysteine, and L-methionine increased the CO_2 uptake of water by about 20%, while phenylalanine and threonine had negligible promotion or inhibition effects on CO_2 hydrate at 0.5 wt% in both modes. This demonstrated that valine is capable of promoting both CH_4 and CO_2 hydrate formation.[56,66]

A follow-up study by Prasad and Kiran[64] examined methionine and phenylalanine on CH_4, CO_2, and their mixtures at 0.5 wt% using a nonstirred and isochoric mode. The study reported low hydrate conversion efficiency for phenylalanine in CO_2 hydrate, but both methionine and phenylalanine showed significant hydrate conversion efficiency in CH_4 and mixed $CH_4 + CO_2$ systems. The presence of methionine and phenylalanine enhanced the kinetics of hydrate formation, achieving about 90% gas-to-hydrate conversion and over 85% water-to-hydrate conversion within an hour. Methionine promoted hydrate formation more effectively than phenylalanine in both gas systems, but phenylalanine is more recommended for methane hydrates only.

These findings confirm those of Sa et al.,[80] indicating that amino acids form structure I hydrates. This discovery highlights the potential for amino acids in the separation of CH_4 gas from $CO_2 + CH_4$ gas mixtures and in natural gas storage applications.

3.3.4 Amino Acid Kinetic Promotion Mechanism

The amino acid hydrate promotion mechanism is influenced by numerous factors that are not yet fully understood.[85] Authors speculate that the proposed hydrate promotion effect of amino acids arises from their surface activity and surface adsorption behavior via capillary action.[84,85,87] The surface activity of amino acids, resulting in hydrate formation enhancement, is analogous to that of conventional surfactants. The molecular structure of most amino acids comprises both hydrophilic and hydrophobic components, due to the presence of amine and carboxylic acid groups, as well as diverse side chains. In addition, the side chains of amino acids may vary based on their polarity, charge, and structure. This amphiphilic nature allows amino acids to act as surfactants. For example, leucine, one of the most effective amino acid promoters, has hydrophilic amine and carboxylic acid groups, along with a hydrophobic aliphatic isobutyl side chain. In addition, some amino acids, such as arginine and valine, function as bio-surfactants and suppress protein aggregation.[88–90]

This surfactant behavior enables these amino acids to prevent or break the formation and agglomeration of hydrate nucleus crystals at the gas/liquid interface, thereby allowing more gas to dissolve in the liquid phase and enhancing hydrate gas uptake.

Linga's lab demonstrated that hydrates formed in amino acid solutions are highly flexible and porous, contributing to their hydrate promotion effect.[86] The presence of these porous and flexible hydrates increases the surface adsorption capacity at the gas/liquid interface. This increased surface adsorption facilitates the capillary effect, enabling more liquid to be drawn to the gas/liquid interface and resulting in higher gas uptake during hydrate formation.

It is important to note that the promotion or inhibition mechanism of amino acids in CO_2 systems is partly influenced or controlled by the reaction between amino acids and CO_2 molecules. Zhang et al.[91] provide a comprehensive summary of the reaction between amino acids and CO_2. The zwitterionic reaction mechanism is primarily observed in these interactions. In this process, the amine group of the amino acids initially reacts with CO_2 to form intermediates known as zwitterions. The presence of any base (such as additional amine groups or water) in the system will result in the formation of amino acid salts through a reaction between the zwitterions and the base.[91] Generally, the rate constant of this reaction describes the CO_2 adsorption rate, which is related to the CO_2 hydrate formation rate and uptake. Therefore, amino acids with a fast reaction rate will potentially promote hydrate formation, while those with a slower reaction rate may inhibit it.

3.4 COMPARISON OF AMINO ACIDS WITH OTHER HYDRATE-BASED APPLICATION ADDITIVES

In this section, the thermodynamic and kinetic inhibition or promotion effects of amino acids are compared with those of commercially available inhibitors and promoters to assess their efficiency and applicability in industrial operations. The discussion is organized into two parts: Thermodynamics and Kinetics. Previous studies on hydrate phase behavior in amino acids have not demonstrated a promotion effect on hydrate formation. Consequently, only the thermodynamic inhibition effect is examined in this study. The thermodynamic inhibition effect of the most effective amino acids is compared with commonly used inhibitors such as methanol,[92] ethanol,[93,94] ethylene glycol,[93] diethylene glycol,[93] triethylene glycol,[93] and glycerol[3,95] for methane and carbon dioxide that hydrate at 10 wt%, as illustrated in Figure 3.9.

As shown in Figure 3.9, methanol, ethanol, and ethylene glycol are more efficient than amino acids (glycine and alanine). However, amino acids are environmentally friendly and cost-effective when produced in large quantities. Although amino acids are less efficient than triethylene glycol, they exhibit comparable inhibition performance to glycerol and diethylene glycol in methane and carbon dioxide systems. Therefore, hydrate prevention techniques employing glycerol, diethylene glycol, and triethylene glycol can be substituted with amino acids due to their efficiency and environmental benefits. Nevertheless, the limited solubility of amino acids at high concentrations may pose a challenge for their application in large-scale operations. Notably, proline has demonstrated high solubility in water, making it suitable for hydrate mitigation applications[58].

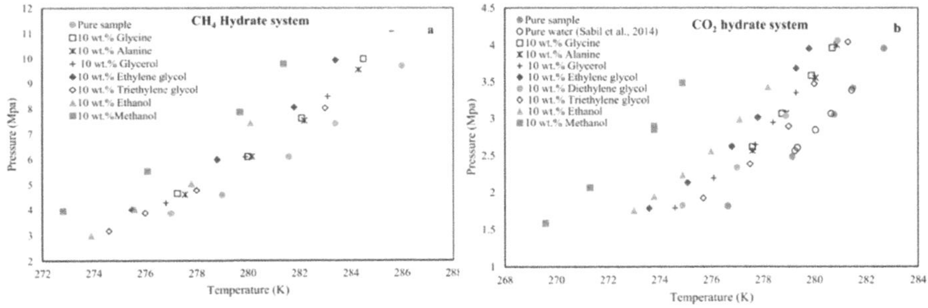

FIGURE 3.9 Comparison of the THI efficiency of amino acids and conventional additives for CH_4 and CO_2 hydrate system at 10 wt%; the pure water data for CH_4 and CO_2 are taken from reference (Bavoh et al., 2016b, 2017; Nasir et al., 2014); (a) CH_4 hydrate system; (b) CO_2 hydrate system.

3.5 MODELING AND SIMULATION OF GAS HYDRATE IN THE PRESENCE OF AMINO ACIDS

Currently, the literature[54–56] has explored thermodynamic modeling of gas hydrate inhibition in amino acids using the Dickens and Quinby-Hunt[96] model, which extends the non-electrolyte hydrate inhibitors model proposed by Pieroen[107]. This model assumes that amino acids exhibit salt-like behavior, allowing for the adaptation of existing gas hydrate models developed for salts to amino acids. For detailed model formulations and assumptions, refer to the works of Bavoh et al.,[54,55] Dickens and Quinby-Hunt,[96] and Pieroen[107]. The simplified form of the model is presented in equation (3.1):

$$\left[\frac{1}{T_w}-\frac{1}{T_{aa}}\right]=\frac{n\Delta H_{FUS(i)}}{\Delta H_d}\left[\frac{1}{T_{f(i)}}-\frac{1}{T_{fa}}\right]$$

where $T_{f(i)}$ and T_{fa} are the freezing point temperatures of water (at 273.15 K) and the water + amino acid solution, $\Delta H_{FUS(i)}$ is the heat of fusion of ice (6008 J/mol), ΔH_d is the molar enthalpy of dissociation of the gas system (which can be determined experimentally or via the Clausius-Clapeyron equation), n is the hydration number of the gas system obtainable from experimental data or literature sources,[97] R is the universal gas constant, and T_w and T_{aa} are the hydrate phase boundary temperatures in pure water and the water + amino acid solution, respectively. The model accurately predicts hydrate phase boundary conditions for methane and carbon dioxide with an average absolute error (AAE) of less than 0.2 K.[54,55]

Naeiji et al.[73] and Rad et al.[74] modeled the formation rate of THF and ethane hydrate, employing the thermodynamic natural path in a constant volume process. Roosta et al.[77] subsequently modeled the kinetic impact of amino acids on CO_2 hydrates using a chemical affinity model. Their parameters concurred with experimental results, indicating that the presence of amino acids decreases the rate of CO_2 hydrate formation. Moreover, a molecular dynamics (MD) simulation study on

CH_4 hydrates was conducted by Oluwunmi et al.,[98] suggesting that asparagine inhibits hydrate formation and growth by adsorbing at the water/methane interface due to its hydrophilic nature. In addition, Bhattacharjee et al.[71] simulated CH_4 hydrate formation in the presence of histidine, finding that it promotes CH_4 hydrate formation, corroborating experimental results. A recent MD simulation by Maddah et al.[99] on the methane hydrate inhibition impact of glycine, proline, serine, and alanine confirmed their KHI behavior. The study evaluated parameters including the radial distribution function, four-body structural order parameter, potential energy, mean square displacement, density, and hydrogen bond formation. The findings indicated that the instability of the structure I gas hydrate, responsible for methane hydrate inhibition, is attributed to van der Waals interactions, potential energy, and electrostatic forces between each amino acid and water molecules in the solution. The Conductor-like Screening Model for Real Solvents (COSMO-RS) software,[69,100,101,102] recognized for its effectiveness and speed in screening compounds/additives, has been proposed as a potent tool for screening amino acids in gas hydrate studies through hydrogen bonding energies and sigma profile/potential predictions.[54,55]

3.6 RECOMMENDATIONS

Amino acids have shown significant potential for efficiency in various gas hydrate-based technologies, potentially leading to commercialization. Although they exhibit weaknesses in thermodynamically promoting hydrate formation, they possess substantial hydrate thermodynamic and kinetic inhibition potentials and are highly effective in kinetically promoting hydrate formation for natural gas storage, CO_2 capture, and gas separation. In addition, amino acids are relatively inexpensive, biodegradable, environmentally friendly, noncorrosive, and do not produce foams, making them promising for future industrial applications in gas hydrate-based technologies. However, to effectively utilize amino acids, their hydrate inhibition and promotion efficiency must be enhanced to meet industrial standards. The amino acids currently studied do not inhibit and promote gas hydrate formation as effectively as the conventional additives used in the industry. Therefore, research toward the commercialization of amino acids in hydrate-based technology should focus on the following:

- Concentration is highly recommended for industrial applications. In addition, some hydrate inhibitors are incompatible with other industrial additives, such as corrosion inhibitors.[11,103,104] Their application affects the performance of such additives, making it important to perform compatibility tests of amino acids with other industrial additives, along with an economic analysis. This is crucial for paving the way for the successful application of amino acids in gas hydrate-based technologies.

3.7 CONCLUSION

The influence of amino acids on gas hydrate formation has been reviewed based on available data in the open literature. The review concludes that most amino acids promote hydrate formation kinetics, while a few, such as glycine and alanine, inhibit gas hydrate formation both thermodynamically and kinetically, thus acting as dual functional inhibitors, similar to ILs. Generally, amino acids are THIs with no reported thermodynamic promotion. The promotion or inhibition effect of amino acids greatly depends on their respective side chain properties (hydropathy, side chain alkyl length, polarity, functional group, etc.), solubility, concentration, studied concentration units, interaction with the guest molecule, and hydrogen bond and electrostatic force of attraction with water molecules. However, the hydropathy of amino acids is less understood, making it difficult to correlate available hydropathy scales with gas hydrate inhibition impact. While amino acids are generally gas hydrate kinetic promoters, some slightly inhibit gas hydrate formation kinetically by perturbing the local water structure and causing lattice distortion and expansion by incorporation into hydrate lattice crystals. Furthermore, the effect of amino acids on hydrate structure characterization is needed for thermodynamic and kinetic modeling purposes. Finally, more MD simulations are needed to understand the gas hydrate inhibition mechanism of amino acids. Amino acids are potential additives for future hydrate-based applications, especially in CO_2 capture and storage and natural gas storage.

REFERENCES

1. Koh CA, Sloan ED, Sum AK, Wu DT. Fundamentals and applications of gas hydrates. *Annu Rev Chem Biomol Eng*. 2011;2:237–257. doi:10.1146/annurev-chembioeng-061010-114152
2. Sloan Jr., E.D., Koh, C.A., & Koh, C.A. Clathrate Hydrates of Natural Gases (3rd ed.). CRC Press. 2007. https://doi.org/10.1201/9781420008494.
3. Breland E, Englezos P. Equilibrium hydrate formation data for carbon dioxide in aqueous glycerol solutions. *J Chem Eng Data*. 1996;41(1):11–13. doi:10.1021/je950181y
4. Babu P, Linga P, Kumar R, Englezos P. A review of the hydrate based gas separation (HBGS) process forcarbon dioxide pre-combustion capture. *Energy*. 2015;85:261–279. doi:10.1016/j.energy.2015.03.103
5. Park S, Lee S, Lee Y, Seo Y. CO2 capture from simulated fuel gas mixtures using semiclathrate hydrates formed by quaternary ammonium salts. *Environ Sci Technol*. 2013;47(13):7571–7577. doi:10.1021/es400966x
6. Lang X, Fan S, Wang Y. Intensification of methane and hydrogen storage in clathrate hydrate and future prospect. *J Nat Gas Chem*. 2010;19(3):203–209. doi:10.1016/S1003-9953(09)60079-7
7. Najibi H, Rezaei R, Javanmardi J, Nasrifar K, Moshfeghian M. Economic evaluation of natural gas transportation from Iran's South-Pars gas field to market. *Appl Therm Eng*. 2009;29(10):2009–2015. doi:10.1016/j.applthermaleng.2008.10.008
8. Strobel TA, Taylor CJ, Hester KC, Dec SF, Koh CA, Miller KT, Sloan ED. Molecular hydrogen storage in binary THF-H2 clathrate hydrates. *J Phys Chem B*. 2006;110(34):17121–17125. doi:10.1021/jp062139n
9. Tariq M, Rooney D, Othman E, Aparicio S, Atilhan M, Khraisheh M. Gas hydrate inhibition: A review of the role of ionic liquids. *Ind Eng Chem Res*. 2014;53(46):17855–17868. doi:10.1021/ie503559k
10. Tariq M, Rooney D, Othman E, Aparicio S, Atilhan M, Khraisheh M. Gas hydrate inhibition: A review of the role of ionic liquids. *Ind Eng Chem Res*. 2014;53(46):17855–17868. doi:10.1021/ie503559k
11. Kamal MS, Hussein IA, Sultan AS, Von Solms N. Application of various water soluble polymers in gas hydrate inhibition. *Renew Sustain Energy Rev*. 2016;60:206–225. doi:10.1016/j.rser.2016.01.092

12. Chong ZR, Chan AHM, Babu P, Yang M, Linga P. Effect of NaCl on methane hydrate formation and dissociation in porous media. *J Nat Gas Sci Eng.* 2015;27:178–189. doi:10.1016/j.jngse.2015.08.055
13. Nashed O, Partoon B, Lal B, Sabil KM, Shariff AM. Review the impact of nanoparticles on the thermodynamics and kinetics of gas hydrate formation. *J Nat Gas Sci Eng.* 2018;55:452–465. doi:10.1016/j.jngse.2018.05.022
14. Pan Z, Liu Z, Zhang Z, Shang L, Ma S. Effect of silica sand size and saturation on methane hydrate formation in the presence of SDS. *J Nat Gas Sci Eng.* 2018;56(February): 266–280. doi:10.1016/j.jngse.2018.06.018
15. Liu Z, Pan Z, Zhang Z, Liu P, Shang L, Li B. Effect of porous media and sodium dodecyl sulphate complex system on methane hydrate formation. *Energy and Fuels.* 2018;32(5):5736–5749. doi:10.1021/acs.energyfuels.8b00041
16. Li S, Fan S, Wang J, Lang X, Wang Y. Clathrate hydrate capture of CO_2 from simulated flue gas with cyclopentane/water emulsion. *Chin J Chem Eng.* 2010;18(2):202–206. doi:10.1016/S1004-9541(08)60343-2
17. Hao W, Wang J, Fan S, Hao W. Evaluation and analysis method for natural gas hydrate storage and transportation processes. *Energy Convers Manag.* 2008;49(10):2546–2553. doi:10.1016/j.enconman.2008.05.016
18. Veluswamy HP, Kumar A, Seo Y, Lee JD, Linga P. A review of solidified natural gas (SNG) technology for gas storage via clathrate hydrates. *Appl Energy.* 2018;216 (February):262–285. doi:10.1016/j.apenergy.2018.02.059
19. Khan MS, Bavoh CB, Partoon B, Lal B, Bustam MA, Shariff AM. Thermodynamic effect of ammonium based ionic liquids on CO_2 hydrates phase boundary. *J Mol Liq.* 2017;238:533–539. doi:10.1016/j.molliq.2017.05.045
20. Bavoh CB, Nashed O, Khan MS, Partoon B, Lal B, Sharif AM. The impact of amino acids on methane hydrate phase boundary and formation kinetics. *J Chem Thermodyn.* 2018;117:48–53. doi:10.1016/j.jct.2017.09.001
21. Xiao C, Adidharma H. Dual function inhibitors for methane hydrate. *Chem Eng Sci.* 2009;64(7):1522–1527. doi:10.1016/j.ces.2008.12.031
22. Khan MS, Lal B, Keong LK, Ahmed I. Tetramethyl ammonium chloride as dual functional inhibitor for methane and carbon dioxide hydrates. *Fuel.* 2019;236(August 2018):251–263. doi:10.1016/j.fuel.2018.09.001
23. Bavoh CB, Khan MS, Lal B, Bt Abdul Ghaniri NI, Sabil KM. New methane hydrate phase boundary data in the presence of aqueous amino acids. *Fluid Phase Equilib.* 2018;478:129–133. doi:10.1016/j.fluid.2018.09.011
24. Yaqub S, Lal B, Partoon B, Mellon NB. Investigation of the task oriented dual function inhibitors in gas hydrate inhibition: A review. *Fluid Phase Equilib.* 2018;477:40–57. doi:10.1016/j.fluid.2018.08.015
25. Thuy Pham TP, Cho CW, Yun YS. Environmental fate and toxicity of ionic liquids: A review. *Water Res.* 2010;44(2):352–372. doi:10.1016/j.watres.2009.09.030
26. Zare M, Haghtalab A, Ahmadi AN, Nazari K. Experiment and thermodynamic modeling of methane hydrate equilibria in the presence of aqueous imidazolium-based ionic liquid solutions using electrolyte cubic square well equation of state. *Fluid Phase Equilib.* 2013;341:61–69. doi:10.1016/j.fluid.2012.12.024
27. Sa JH, Lee BR, Park DH, Han K, Chun HD, Lee KH. Amino acids as natural inhibitors for hydrate formation in CO_2 sequestration. *Environ Sci Technol.* 2011;45(13):5885–5891. doi:10.1021/es200552c
28. Hecht D, Tadesse L, Walters L. 1993. 9,002-. 2(10):2–3.
29. Pertsemlidis A, Saxena AM, Soper AK, Head-Gordon T, Glaeser RM. Direct evidence for modified solvent structure within the hydration shell of a hydrophobic amino acid. *Proc Natl Acad Sci U S A.* 1996;93(20):10769–10774. doi:10.1073/pnas.93.20.10769

30. Zheng J, Zhang P, Linga P. Semiclathrate hydrate process for pre-combustion capture of CO_2 at near ambient temperatures. *Appl Energy [Internet]*. 2017;194:267–278. doi:10.1016/j.apenergy.2016.10.118
31. Madeira PP, Bessa A, Álvares-Ribeiro L, Raquel Aires-Barros M, Rodrigues AE, Uversky VN, Zaslavsky BY. Amino acid/water interactions study: A new amino acid scale. *J Biomol Struct Dyn*. 2014;32(6):959–968. doi:10.1080/07391102.2013.800994
32. Vaitheeswaran S, Thirumalai D. 2008. Interactions between Amino acid side chains in cylindrical hydrophobic nanopores with applications to peptide stability. [place unknown].
33. Badawy WA, Ismail KM, Fathi AM. Effect of Ni content on the corrosion behavior of Cu-Ni alloys in neutral chloride solutions. *Electrochim Acta*. 2005;50(18):3603–3608. doi:10.1016/j.electacta.2004.12.030
34. Barouni K, Bazzi L, Salghi R, Mihit M, Hammouti B, Albourine A, El Issami S. Some amino acids as corrosion inhibitors for copper in nitric acid solution. *Mater Lett*. 2008;62(19):3325–3327. doi:10.1016/j.matlet.2008.02.068
35. Arakawa T, Tsumoto K, Kita Y, Chang B, Ejima D. Biotechnology applications of amino acids in protein purification and formulations. *Amino Acids*. 2007;33(4):587–605. doi:10.1007/s00726-007-0506-3
36. Arakawa T, Ejima D, Tsumoto K, Obeyama N, Tanaka Y, Kita Y, Timasheff SN. Suppression of protein interactions by arginine: A proposed mechanism of the arginine effects. *Biophys Chem*. 2007;127(1–2):1–8. doi:10.1016/j.bpc.2006.12.007
37. Sa JH, Lee BR, Park DH, Han K, Chun HD, Lee KH. Amino acids as natural inhibitors for hydrate formation in CO_2 sequestration. *Environ Sci Technol*. 2011;45(13):5885–5891. doi:10.1021/es200552c
38. Kakati H, Mandal A, Laik S. Promoting effect of Al_2O_3/ZnO-based nanofluids stabilized by SDS surfactant on CH_4+C_2H_6+C_3H_8 hydrate formation. *J Ind Eng Chem [Internet]*. 2016;35:357–368. doi:10.1016/j.jiec.2016.01.014
39. Talaghat MR. Experimental investigation of induction time for double gas hydrate formation in the simultaneous presence of the PVP and l-tyrosine as kinetic inhibitors in a mini flow loop apparatus. *J Nat Gas Sci Eng [Internet]*. 2014;19:215–220. doi:10.1016/j.jngse.2014.05.010
40. Sa JH, Kwak GH, Han K, Ahn D, Cho SJ, Lee JD, Lee KH. Inhibition of methane and natural gas hydrate formation by altering the structure of water with amino acids. *Sci Rep [Internet]*. 2016;6(April):1–9. doi:10.1038/srep31582
41. Roosta H, Dashti A, Mazloumi SH, Varaminian F. Inhibition properties of new amino acids for prevention of hydrate formation in carbon dioxide-water system: Experimental and modeling investigations. *J Mol Liq [Internet]*. 2016;215:656–663. doi:10.1016/j.molliq.2016.01.039
42. Bavoh CB, Lal B, Nashed O, Khan MS, Keong LK, Bustam MA. COSMO-RS: An ionic liquid prescreening tool for gas hydrate mitigation. *Chin J Chem Eng [Internet]*. 2016;24(11):1619–1624. doi:10.1016/j.cjche.2016.07.014
43. Bavoh CB, Partoon B, Lal B, Kok Keong L. Methane hydrate-liquid-vapour-equilibrium phase condition measurements in the presence of natural amino acids. *J Nat Gas Sci Eng [Internet]*. 2017;37:425–434. doi:10.1016/j.jngse.2016.11.061
44. Veluswamy HP, Hong QW, Linga P. Morphology study of methane hydrate formation and dissociation in the presence of amino acid. *Cryst Growth Des*. 2016;16(10):5932–5945. doi:10.1021/acs.cgd.6b00997
45. Cai Y, Chen Y, Li Q, Li L, Huang H, Wang S, Wang W. CO_2 Hydrate formation promoted by a natural amino acid l-methionine for possible application to CO_2 capture and storage. *Energy Technology*. 2017;5(8):1195–1199. doi:10.1002/ente.201600731
46. Veluswamy HP, Lee PY, Premasinghe K, Linga P. Effect of biofriendly amino acids on the kinetics of methane hydrate formation and dissociation. *Ind Eng Chem Res*. 2017;56(21):6145–6154. doi:10.1021/acs.iecr.7b00427

47. Mannar N, Bavoh CB, Baharudin AH, Lal B, Mellon NB. Thermophysical properties of aqueous lysine and its inhibition influence on methane and carbon dioxide hydrate phase boundary condition. *Fluid Phase Equilib [Internet].* 2017;454:57–63. doi:10.1016/j.fluid.2017.09.012
48. Bavoh CB, Khan MS, Ting VJ, Lal B, Ofei TN, Ben-Awuah J, Ayoub M, Shariff ABM. The effect of acidic gases and thermodynamic inhibitors on the hydrates phase boundary of synthetic malaysia natural gas. *IOP Conf Ser Mater Sci Eng.* 2018;458(1). doi:10.1088/1757-899X/458/1/012016
49. Prasad PSR, Kiran BS. Are the amino acids thermodynamic inhibitors or kinetic promoters for carbon dioxide hydrates? *J Nat Gas Sci Eng [Internet].* 2018;52:461–466. doi:10.1016/j.jngse.2018.02.001
50. Liu Y, Chen B, Chen Y, Zhang S, Guo W, Cai Y, Tan B, Wang W. Methane storage in a hydrated form as promoted by leucines for possible application to natural gas transportation and storage. *Energy Technol.* 2015;3(8):815–819. doi:10.1002/ente.201500048
51. Khan MS, Liew CS, Kurnia KA, Cornelius B, Lal B. Application of COSMO-RS in investigating ionic liquid as thermodynamic hydrate inhibitor for methane hydrate. *Procedia Eng [Internet].* 2016;148:862–869. doi:10.1016/j.proeng.2016.06.452
52. Long Z, Zhou X, He Y, Li D, Liang D. Performance of mixture of ethylene glycol and glycine in inhibiting methane hydrate formation. *J Nat Gas Sci Eng [Internet].* 2018;56:134–140. doi:10.1016/j.jngse.2018.05.034
53. Bavoh CB, Nashed O, Khan MS, Partoon B, Lal B, Sharif AM. The impact of amino acids on methane hydrate phase boundary and formation kinetics. *J Chem Thermodyn [Internet].* 2018;117:48–53. doi:10.1016/j.jct.2017.09.001
54. Bavoh CB, Partoon B, Lal B, Gonfa G, Foo Khor S, Sharif AM. Inhibition effect of amino acids on carbon dioxide hydrate. *Chem Eng Sci.* 2017;171:331–339. doi:10.1016/j.ces.2017.05.046
55. Bavoh CB, Partoon B, Lal B, Kok Keong L. Methane hydrate-liquid-vapour-equilibrium phase condition measurements in the presence of natural amino acids. *J Nat Gas Sci Eng.* 2017;37:425–434. doi:10.1016/j.jngse.2016.11.061
56. Bavoh CB, Khan MS, Ting VJ, Lal B, Ofei TN, Ben-Awuah J, Ayoub M, Shariff ABM. The effect of acidic gases and thermodynamic inhibitors on the hydrates phase boundary of synthetic Malaysia natural gas. *IOP Conf Ser Mater Sci Eng.* 2018;458(1). doi:10.1088/1757-899X/458/1/012016
57. Mannar N, Bavoh CB, Baharudin AH, Lal B, Mellon NB. Thermophysical properties of aqueous lysine and its inhibition influence on methane and carbon dioxide hydrate phase boundary condition. *Fluid Phase Equilib.* 2017;454:57–63. doi:10.1016/j.fluid.2017.09.012
58. Sa JH, Kwak GH, Han K, Ahn D, Cho SJ, Lee JD, Lee KH. Inhibition of methane and natural gas hydrate formation by altering the structure of water with amino acids. *Sci Rep.* 2016;6(April):1–9. doi:10.1038/srep31582
59. Bavoh CB, Lal B, Osei H, Sabil KM, Mukhtar H. A review on the role of amino acids in gas hydrate inhibition, CO_2 capture and sequestration, and natural gas storage. *J Nat Gas Sci Eng.* 2019;64:52–71. doi:10.1016/j.jngse.2019.01.020
60. Mech D, Pandey G, Sangwai JS. Effect of molecular weight of polyethylene glycol on the equilibrium dissociation pressures of methane hydrate system. *J Chem Eng Data.* 2015;60(6):1878–1885. doi:10.1021/acs.jced.5b00088
61. Yousif MH. Effect of underinhibition with methanol and ethylene glycol on the hydrate-control process. *SPE Prod and Facil.* 1998;13(3):184–189. doi:10.2118/50972-PA
62. Bavoh CB, Partoon B, Lal B, Kok Keong L. Methane hydrate-liquid-vapour-equilibrium phase condition measurements in the presence of natural amino acids. *J Nat Gas Sci Eng.* 2017;37:425–434. doi:10.1016/j.jngse.2016.11.061

63. Long Z, Zhou X, He Y, Li D, Liang D. Performance of mixture of ethylene glycol and glycine in inhibiting methane hydrate formation. *J Nat Gas Sci Eng*. 2018;56:134–140. doi:10.1016/j.jngse.2018.05.034
64. Prasad PSR, Sai Kiran B. Clathrate hydrates of greenhouse gases in the presence of natural amino acids: Storage, transportation and separation applications. *Sci Rep*. 2018;8(1):1–10. doi:10.1038/s41598-018-26916-1
65. Sa JH, Kwak GH, Han K, Ahn D, Lee KH. Gas hydrate inhibition by perturbation of liquid water structure. *Sci Rep*. 2015;5:1–9. doi:10.1038/srep11526
66. Prasad PSR, Kiran BS. Are the amino acids thermodynamic inhibitors or kinetic promoters for carbon dioxide hydrates? *J Nat Gas Sci Eng*. 2018;52:461–466. doi:10.1016/j.jngse.2018.02.001
67. Zimmerman JM, Eliezer N, Simha R. The characterization of amino acid sequences in proteins by statistical methods. *J Theor Biol*. 1968;21(2):170–201. doi:10.1016/0022-5193(68)90069-6
68. Naeiji P, Mottahedin M, Varaminian F. Separation of methane-ethane gas mixtures via gas hydrate formation. *Sep Purif Technol*. 2014;123:139–144. doi:10.1016/j.seppur.2013.12.028
69. Bavoh CB, Lal B, Nashed O, Khan MS, Keong LK, Bustam MA. COSMO-RS: An ionic liquid prescreening tool for gas hydrate mitigation. *Chin J Chem Eng*. 2016;24(11):1619–1624. doi:10.1016/j.cjche.2016.07.014
70. Kyte J, Doolittle RF. 1982. A simple method for displaying the hydropathic character of a protein. *J. Mol. Biol*. 1982;157:105–132.
71. Bhattacharjee G, Choudhary N, Kumar A, Chakrabarty S, Kumar R. Effect of the amino acid L-histidine on methane hydrate growth kinetics. *J Nat Gas Sci Eng*. 2016;35:1453–1462. doi:10.1016/j.jngse.2016.05.052
72. Kakati H, Mandal A, Laik S. Synergistic effect of polyvinylpyrrolidone (PVP) and l-tyrosine on kinetic inhibition of $CH_4 + C_2H_4 + C_3H_8$ hydrate formation. *J Nat Gas Sci Eng*. 2016;34:1361–1368. doi:10.1016/j.jngse.2016.08.027
73. Naeiji P, Arjomandi A, Varaminian F. Amino acids as kinetic inhibitors for tetrahydrofuran hydrate formation: Experimental study and kinetic modeling. *J Nat Gas Sci Eng*. 2014;21:64–70. doi:10.1016/j.jngse.2014.07.029
74. Abbasian Rad S, Rostami Khodaverdiloo K, Karamoddin M, Varaminian F, Peyvandi K. Kinetic study of amino acids inhibition potential of glycine and l-leucine on the ethane hydrate formation. *J Nat Gas Sci Eng*. 2015;26:819–826. doi:10.1016/j.jngse.2015.06.053
75. Talaghat MR. Experimental investigation of induction time for double gas hydrate formation in the simultaneous presence of the PVP and l-tyrosine as kinetic inhibitors in a mini flow loop apparatus. *J Nat Gas Sci Eng*. 2014;19:215–220. doi:10.1016/j.jngse.2014.05.010
76. Perfeldt CM, Chua PC, Daraboina N, Friis D, Kristiansen E, Ramløv H, Woodley JM, Kelland MA, Von Solms N. Inhibition of gas hydrate nucleation and growth: Efficacy of an antifreeze protein from the longhorn beetle *Rhagium mordax*. *Energy and Fuels*. 2014;28(6):3666–3672. doi:10.1021/ef500349w
77. Roosta H, Dashti A, Mazloumi SH, Varaminian F. Inhibition properties of new amino acids for prevention of hydrate formation in carbon dioxide-water system: Experimental and modeling investigations. *J Mol Liq*. 2016;215:656–663. doi:10.1016/j.molliq.2016.01.039
78. Sa JH, Kwak GH, Lee BR, Park DH, Han K, Lee KH. Hydrophobic amino acids as a new class of kinetic inhibitors for gas hydrate formation. *Sci Rep*. 2013;3. doi:10.1038/srep02428

79. Xu S, Fan S, Fang S, Wang Y, Lang X. Excellent synergy effect on preventing CH4 hydrate formation when glycine meets polyvinylcaprolactam. *Fuel*. 2017;206:19–26. doi:10.1016/j.fuel.2017.05.030
80. Sa JH, Kwak GH, Lee BR, Ahn D, Lee KH. Abnormal incorporation of amino acids into the gas hydrate crystal lattice. *Phys Chem Chem Phys*. 2014;16(48):26730–26734. doi:10.1039/c4cp05056h
81. Sefidroodi H, Cheng Chua P, Kelland MA. THF hydrate crystal growth inhibition with small anionic organic compounds and their synergistic properties with the kinetic hydrate inhibitor poly(N-vinylcaprolactam). *Chem Eng Sci*. 2011;66(10):2050–2056. doi:10.1016/j.ces.2011.01.014
82. Sowa B, Zhang XH, Hartley PG, Dunstan DE, Kozielski KA, Maeda N. Formation of ice, tetrahydrofuran hydrate, and methane/propane mixed gas hydrates in strong monovalent salt solutions. *Energy and Fuels*. 2014;28(11):6877–6888. https://doi.org/10.1021/ef501701y
83. Kakati H, Mandal A, Laik S. Promoting effect of Al_2O_3/ZnO-based nanofluids stabilized by SDS surfactant on CH_4+C_2H_6+C_3H_8 hydrate formation. *J Ind Eng Chem*. 2016; 35:357–368. doi:10.1016/j.jiec.2016.01.014
84. Veluswamy HP, Lee PY, Premasinghe K, Linga P. Effect of biofriendly amino acids on the kinetics of methane hydrate formation and dissociation. *Ind Eng Chem Res*. 2017;56(21):6145–6154. doi:10.1021/acs.iecr.7b00427
85. Liu Y, Chen B, Chen Y, Zhang S, Guo W, Cai Y, Tan B, Wang W. Methane storage in a hydrated form as promoted by leucines for possible application to natural gas transportation and storage. *Energy Technol*. 2015;3(8):815–819. doi:10.1002/ente.201500048
86. Veluswamy HP, Hong QW, Linga P. Morphology study of methane hydrate formation and dissociation in the presence of amino acid. *Cryst Growth Des*. 2016;16(10):5932–5945. doi:10.1021/acs.cgd.6b00997
87. Cai Y, Chen Y, Li Q, Li L, Huang H, Wang S, Wang W. CO_2 hydrate formation promoted by a natural amino acid l-methionine for possible application to CO_2 capture and storage. *Energy Technol*. 2017;5(8):1195–1199. doi:10.1002/ente.201600731
88. Infante MR, Pérez L, Pinazo A, Clapés P, Morán MC, Angelet M, García MT, Vinardell MP. Amino acid-based surfactants. *Comptes Rendus Chimie*. 2004;7(6–7):583–592. doi:10.1016/j.crci.2004.02.009
89. Infante MR, Pinazo A, Seguer J. Non-conventional surfactants from amino acids and glycolipids: Structure, preparation and properties. *Colloids Surf A Physicochem Eng Asp*. 1997;123–124(Table 1):49–70. doi:10.1016/S0927-7757(96)03793-4
90. Pinazo A, Pons R, Pérez L, Infante MR. Amino acids as raw material for biocompatible surfactants. *Ind Eng Chem Res*. 2011;50(9):4805–4817. doi:10.1021/ie1014348
91. Zhang Z, Li Y, Zhang W, Wang J, Soltanian MR, Olabi AG. Effectiveness of amino acid salt solutions in capturing CO_2: A review. *Renew Sustain Energy Rev*. 2018;98:179–188. doi:10.1016/j.rser.2018.09.019
92. Ng HJ, Robinson DB. Hydrate formation in systems containing methane, ethane, propane, carbon dioxide or hydrogen sulfide in the presence of methanol. *Fluid Phase Equilib*. 1985;21(1–2):145–155. doi:10.1016/0378-3812(85)90065-2
93. Maekawa T. Equilibrium conditions of xenon hydrates in the presence of aqueous solutions of alcohols, glycols, and glycerol. *J Chem Eng Data*. 2016;61(1):662–665. doi:10.1021/acs.jced.5b00791
94. Mohammadi AH, Afzal W, Richon D. Experimental data and predictions of dissociation conditions for ethane and propane simple hydrates in the presence of distilled water and methane, ethane, propane, and carbon dioxide simple hydrates in the presence of ethanol aqueous solutions. *J Chem Eng Data*. 2008;53(1):73–76. doi:10.1021/je700383p

95. Mohammadi AH, Kraouti I, Richon D. Experimental data and predictions of dissociation conditions for methane, ethane, propane, and carbon dioxide simple hydrates in the presence of glycerol aqueous solutions. *Ind Eng Chem Res.* 2008;47(21):8492–8495. doi:10.1021/ie801018e
96. Dickens GR, Quinby-Hunt MS. Methane hydrate stability in pore water: A simple theoretical approach for geophysical applications. *J Geophys Res Solid Earth.* 1997;102(B1): 773–783. doi:10.1029/96jb02941
97. Anderson GK. Enthalpy of dissociation and hydration number of methane hydrate from the Clapeyron equation. *J Chem Thermodyn.* 2004;36(12):1119–1127. doi:10.1016/j.jct.2004.07.005
98. Oluwunmi PA, Finney AR, Rodger PM. Molecular dynamics screening for new kinetic inhibitors of methane hydrate. *Can J Chem.* 2015;93(9):1043–1049. doi:10.1139/cjc-2015-0003
99. Maddah M, Maddah M, Peyvandi K. Molecular dynamics simulation of methane hydrate formation in presence and absence of amino acid inhibitors. *J Mol Liq.* 2018;269:721–732. doi:10.1016/j.molliq.2018.08.108
100. Khan MS, Liew CS, Kurnia KA, Cornelius B, Lal B. Application of COSMO-RS in investigating ionic liquid as thermodynamic hydrate inhibitor for methane hydrate. *Procedia Eng.* 2016;148:862–869. doi:10.1016/j.proeng.2016.06.452
101. Klamt A. COSMO-RS for aqueous solvation and interfaces. *Fluid Phase Equilib.* 2016;407:152–158. doi:10.1016/j.fluid.2015.05.027
102. Klamt A. The COSMO and COSMO-RS solvation models. *Wiley Interdiscip Rev Comput Mol Sci.* 2011;1(5):699–709. doi:10.1002/wcms.56
103. Kelland MA. History of the development of low dosage hydrate inhibitors. *Energy and Fuels.* 2006;20(3):825–847. doi:10.1021/ef050427x
104. Kelland MA, Svartaas TM, Øvsthus J, Namba T. A new class of kinetic hydrate inhibitor. *Ann N Y Acad Sci.* 2000;912(281):281–293. doi:10.1111/j.1749-6632.2000.tb06782.x
105. Roosta H, Dashti A, Mazloumi SH, Varaminian F. The dual effect of amino acids on the nucleation and growth rate of gas hydrate in ethane + water, methane + propane + water and methane + THF + water systems. Fuel [Internet]. 2018;212(August 2016):151–161. doi:10.1016/j.fuel.2017.10.027
106. Wilce MCJ, Aguilar MI, Hearn MTW. Physicochemical Basis of Amino Acid Hydrophobicity Scales: Evaluation of Four New Scales of Amino Acid Hydrophobicity Coefficients Derived from RP-HPLC of Peptides. *Anal Chem.* 1995;67(7):1210–1219. doi:10.1021/ac00103a012
107. Pieroen, A.P., 1955. Gas hydrates- approximate relations between heat formation, composition and equilibrium temperature lowering by inhibitors. *Rec. Trav. Chim.* 1955;74: 955–1002
108. Kumar, R., Roy, S., Bhattacharjee, G., Kumar, A., Choudhary, N., Pande, N.K., Chugh, P., Kashyap, R. A Process for Dissociation of hydrates in the presence of additives of hydrate dissociation promoters. 2017.

4 Amino Acids as Corrosion Inhibitors

Joshua Kojo Aduampong Mantey,
Benjamin Edem Meteku and
Eric Junior Baffoe

4.1 FUNDAMENTALS OF CORROSION

The chemical disintegration of materials, including metals, semiconductors, insulators, and even polymers, brought on by exposure to the environment is known as corrosion. Corrosion is also the environment-induced degradation of desired physicochemical properties in materials. This natural phenomenon can affect both metallic and nonmetallic materials, including plastics, linings, and paints. The surrounding environment may consist of a gas phase, which can be dry or contain moisture, as well as an aqueous or nonaqueous electrolyte. The destructive deterioration of a metal brought on by a chemical or electrochemical reaction with its surroundings is referred to as corrosion. Instead of corrosion, physical deterioration is referred to as erosion, galling, or wear. Physical damage from chemical attacks on businesses can be characterized by corrosive wear, fretting corrosion, and corrosion-erosion. Nonmetals are not covered by this definition of corrosion.

Corrosion processes take place at the surface and can involve various materials in contact with the gas or liquid phase. Historically, the engineering and scientific communities have distinguished between corrosion in electrolytes and corrosion in the gas phase at elevated temperatures, commonly referred to as wet corrosion, hot corrosion, and dry corrosion. Although the apparent impacts change depending on the environment, both professions frequently employ comparable analytic tools in basic research to learn about the underlying mechanisms. Most corrosion processes occur through electrochemical reactions, with their rates largely dependent on operating conditions. Metallic corrosion is categorized into different types based on the underlying mechanism or the environmental factors involved. However, rust formation is expensive and can cause significant damage. To minimize or prevent the harmful effects of corrosion, it is crucial to monitor its progression and adopt effective prevention methods.

In this context, striking a balance between cost-effective and efficient corrosion protection technologies, along with engineering practicality, requires the adoption of innovative approaches.[1] The study of corrosion is an interdisciplinary field that includes engineering, metallurgy, surface science, electrochemistry, physics,

DOI: 10.1201/9781003502388-4

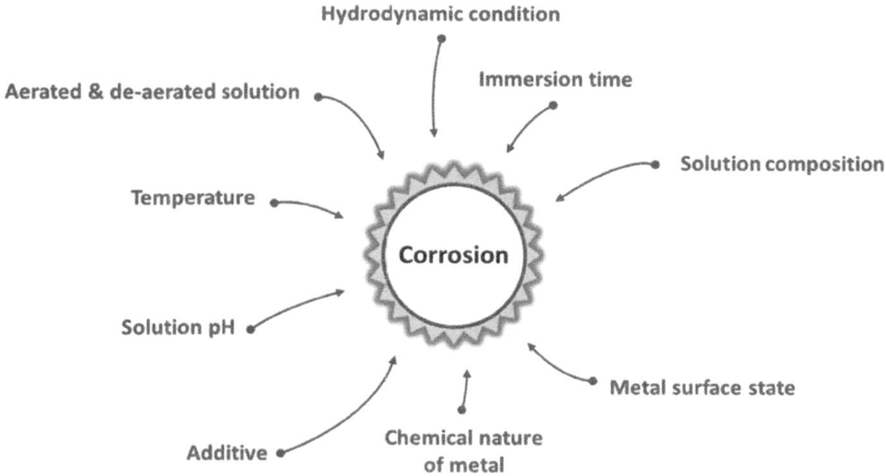

FIGURE 4.1 Some factors affecting the rate of aqueous corrosion of metallic materials. Reproduced with permission from reference 1.

Source: Copyright © 2021, American Chemical Society.

materials science, and theoretical calculations. To examine reaction mechanisms, a variety of in situ and ex situ methods are employed. This chapter will cover these techniques as we examine the various corrosion phenomena and traits. The complexity of the phenomenon necessitates the use of multiple methodologies to gain a deeper understanding. Theorists and experimentalists with expertise in electrochemistry, electrode kinetics, surface science, surface analytical techniques such as x-ray photoelectron spectroscopy, Auger electron spectroscopy, and other techniques utilizing synchrotron radiation in the visible, x-ray, and infrared ranges must therefore work together in groups of experts from different fields.[2] Acidic media (such as hydrochloric acid, sulfuric acid, and nitric acid), alkaline media (like sodium hydroxide and potassium hydroxide), seawater, and saline solutions are examples of corrosive environments. Fontana claims that a metal material reverts to its stable thermodynamic state through the spontaneous process of corrosion. Operating conditions have a major impact on the majority of corrosion processes, which are electrochemical in nature and involve oxidation-reduction reactions. A number of factors have led to an increase in interest in corrosion. For instance, scientists are working to learn more about how newly created materials, like alloys, behave when they corrode.[1] Furthermore, pollution, acidification, and industrial activity are making the atmosphere and water more corrosive (Figure 4.1).

4.2 FACTORS AFFECTING CORROSION

The rate at which metal components corrode is greatly influenced by the flow of corrosive fluids. While higher flow rates for passivating materials can lessen metal deterioration, in certain situations, increased flow of corrosive media over the metal surface speeds up corrosion. The degree and kind of corrosion are largely determined by the state of the metal surface. Deposits or films can act as places where corrosion starts, and biological growths can alter the chemistry of the surface to encourage the dissolution of metal. Furthermore, concentration cells can form on uneven surfaces or in small spaces, hastening corrosion. Surface conditions during corrosion are also influenced by the microstructure of the metal; corrosion behavior is influenced by elements like grain size, orientation, and the presence of secondary phases. The length of time spent in a corrosive environment affects how quickly things corrode. While some corrosion processes move slowly and are less obvious right away, others happen quickly and cause significant damage.

4.3 INTRODUCTION

The term "corrosion" originates from Latin, where *rodere* means "gnawing," and *corrodere* translates to "gnawing to pieces".[1]

Metals are generally found in the form of compounds such as oxides, sulfides, silicates, and carbonates, with only a few metals occurring naturally in their pure state. Compounds are thermodynamically more stable than metals. Extracting a metal from ore is reduced. Reducing iron oxide with carbon produces metallic iron.

$$2\,Fe_2O_3 + 3\,C \rightarrow 4\,Fe + 3\,CO_2;$$

Corrosion is not just a natural occurrence but has significant consequences in three key areas: economics, safety, and environmental harm. Although metallic corrosion may seem harmless, it impacts various industries. The financial burden of corrosion plays a major role in the national economy, with corrosion-related costs typically representing about 2–4% of the gross national product.

As a result, corrosion personnel must implement corrosion control methods to avoid corrosion losses, which account for approximately 25% of total expenditures.[3] Furthermore, corrosion prevention decreases metal degradation, which contributes considerably to resource preservation while causing little environmental damage. Also, corrosion prevention slows metal degradation, which contributes immensely to resource preservation while causing minimal impact on the environment. Corrosion greatly affects safety. Although industrial staff prioritize safety, accidents can still happen despite extensive precautions. In addition to being costly, corrosion poses risks to both human life and overall safety.

4.4 TYPES OF CORROSION AND HOW THEY OCCUR

Corrosion engineering is sometimes seen as a mysterious discipline. However, with a simple understanding of corrosion cells and chemistry, anyone can identify the sources of corrosion problems. Metal corrosion is a natural process driven by the intrinsic tendency of metals to revert to a more stable state, such as an oxide. Once extracted, metal ores must undergo refining and alloying to become functional materials, a process that requires energy. Over time, thermodynamic factors, particularly entropy, promote the gradual corrosion of these metals.

According to G. N. Lewis, any system left undisturbed will naturally progress toward a state of maximum probability. Metals require energy to remain in their refined state, and without external intervention, they will gradually revert to their more stable, naturally occurring compounds.

Polarization curves are commonly used to assess corrosion rates by applying a current to the metal surface. When the applied current causes a positive shift in the metal's potential, it is referred to as anodic polarization, whereas a negative shift is termed cathodic polarization. The extent of polarization reflects how environmental and surface factors influence the rates of anodic and cathodic reactions. Environmental influences, such as metal ion concentration or dissolved oxygen levels in the solution, contribute to concentration polarization, while surface-related factors, including film formation or adsorption, are classified as activation polarization.

A polarization curve illustrates how concentration and activation processes affect the rates at which anodic or cathodic reactions exchange electrons by charting the variation of potential as a function of current. This effectively provides a measure of the corrosion rate by allowing the determination of reaction rates involved in the corrosion process. Corrosion falls into one or more of the following visual categories.

- General or uniform corrosion
- Galvanic corrosion
- Crevice corrosion
- Pitting
- Intergranular corrosion
- Selective leaching
- Erosion corrosion
- Stress corrosion cracking

4.4.1 General or Uniform Corrosion

The most common type of corrosion is general corrosion, sometimes referred to as uniform attack. It thins the material evenly because it happens consistently over the metal's whole exposed surface. Most metal loss is caused by this kind of corrosion. Due to its consistency and predictability, general corrosion is usually the least expensive type of corrosion to control in industrial settings. By subjecting a sample to the environment, corrosion rates can be readily ascertained. Common prevention methods include designing the material to be thicker than necessary to account for corrosion over the product's lifespan or applying protective coatings, such as paint, to the surface.

In general corrosion, electrochemical reactions occurring in aqueous or atmospheric environments progress uniformly across the entire exposed metal surface, leading to significant metal loss in terms of tonnage. The affected surface typically becomes rough and may exhibit a frosted appearance. Although general corrosion is widespread, it is considered less hazardous than other forms, as the lifespan of equipment or structures can be reliably predicted using basic corrosion tests or visual inspections (Figure 4.2).

FIGURE 4.2 Picture of General corrosion. Reproduced with permission from reference 2. Copyright © 2018 Elsevier Inc.

4.4.2 Galvanic Corrosion

The potential difference between two dissimilar metals causes galvanic corrosion, also known as two-metal corrosion. The potential difference that results from electrically connecting these metals either directly or through an electrolyte causes an electron flow, or current, to flow between them. One of the most prevalent types of corrosion is caused by the interaction of a metal with another conductive material in a corrosive environment. It can be seen in a number of places, including when a copper pipe joins a steel pipe at a water main junction, in microelectronic devices that combine various metals and semiconductors, in metal matrix composites that have reinforcing materials like graphite scattered throughout a metal, or on ships where various metal alloys are submerged in water.

In certain instances, galvanic corrosion leads to rapid metal degradation, while in others, the corrosion of one metal helps protect the other, forming the basis for cathodic protection (CP) through sacrificial anodes. Although galvanic corrosion has been extensively studied and is well understood qualitatively, its complexity has historically made quantitative analysis difficult. However, advancements in technology, particularly the widespread use of computers and specialized software, have greatly enhanced the ability to predict and analyze galvanic corrosion. When two dissimilar conductive materials in electrical contact are exposed to an electrolyte, a current known as the galvanic current flows between them. Galvanic corrosion occurs at the anodic member of the metal pair and follows Faraday's law, meaning its extent is directly proportional to the galvanic current. When the metals are coupled, the additional corrosion at the anode is referred to as local corrosion, which may be greater or lesser than the corrosion that would normally occur if the metals were not electrically connected. The difference between local and normal corrosion is termed the difference effect, which can be either positive or negative. In most cases, the galvanic current reduces the overall corrosion rate of the cathodic member, thereby offering CP to that metal (Figure 4.3).

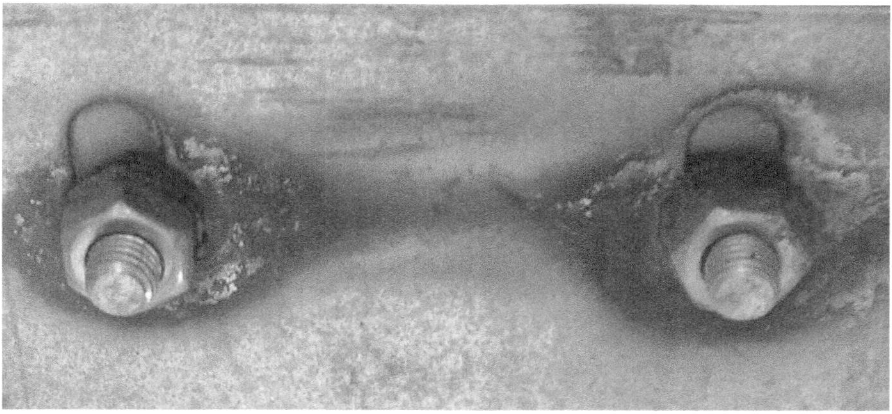

FIGURE 4.3 Light galvanic corrosion around a bolt on a pump housing.[3]

FIGURE 4.4 Failure due to crevice corrosion in the threaded area of a pipe.[3]

4.4.3 Crevice Corrosion

A severe and localized type of corrosion, crevice corrosion, occurs in tight areas like cracks, holes, and crevices, as well as under gaskets, rivets, or bolt heads where stagnant solution builds up. It starts with the development of a corrosion cell, specifically one that concentrates oxygen. The buildup of acidic hydrolyzed salts in the limited area causes the corrosion process to worsen as the oxygen levels inside the crevice drop. An occluded cell that forms beneath a metal surface crevice is the source of crevice corrosion, a particularly dangerous kind of localized corrosion. Areas beneath flanges, gaskets, fasteners, lap joints, disbonded linings or coatings, weld zones, and surface deposits are typical examples. Because the disintegration of the passive layer within the crevice can cause quick and serious damage, materials that depend on passive surface films for corrosion resistance are particularly vulnerable to this type of corrosion (Figure 4.4).

FIGURE 4.5 Hole caused by pitting.[3]

4.4.4 Pitting

Pitting is a highly dangerous form of corrosion that can cause structural failure with minimal overall metal loss. This intensely localized attack leads to rapid penetration through the material's thickness, often resulting in severe damage. The size of pits can vary, from microscopic to relatively large, but their localized nature makes them particularly problematic. What makes pitting especially challenging is that even a small amount of metal loss, concentrated in a single area, can lead to sudden and unexpected failure.

This characteristic makes pitting difficult to detect and prevent, posing significant risks to structures and equipment. Pitting corrosion is an insidious localized form of corrosion that causes extensive damage to structural elements such as stainless steel in chloride environment (Figure 4.5).

4.4.5 Intergranular Corrosion

Intergranular corrosion is a localized attack that occurs along the grain boundaries of metal alloys, especially those with multiple phases or constituents. This form of corrosion originates at the surface and advances through electrochemical activity concentrated near the grain boundaries, creating a narrow path of deterioration. As the grain boundaries corrode, the material can lose its structural integrity, potentially leading to mechanical failure. While the specific mechanisms of intergranular corrosion vary depending on the metal system involved, its microscopic appearance is generally similar across different systems, characterized by the selective attack along the grain boundaries while the bulk grains remain relatively unaffected. This type of corrosion can severely weaken the material, often without obvious visible signs until significant damage has occurred. The effects of this form of attack on mechanical properties may be extremely harmful. Corrosion occurs due to a differential in potential between neighboring grains and the bulk material. The variation in potential could be caused by the difference in chemical composition between the two zones. This can occur when impurities or alloying components migrate to an alloy's grain boundaries. As the concentration of alloying elements in the grain boundary area increases, a second phase or component may separate or precipitate. This may have a different corrosion potential than the grains or matrix, leading to the formation of a local cell. It is a significant concern in materials such as stainless steels, aluminum alloys, and certain nickel-based alloys. Intergranular corrosion can lead to structural weakening, surface pitting, and even catastrophic failure if left unchecked. Understanding its causes, mechanisms, and mitigation strategies is crucial for ensuring the integrity and reliability of metal components in various applications. Grain boundaries are particularly susceptible to corrosion due to their elevated free energy, which makes them prime locations for attack. This vulnerability is further aggravated by the accumulation of impurities at these boundaries and the formation of second-phase particles, which, if more noble, can diminish the corrosion resistance of the adjacent matrix. Consequently, localized corrosion is more likely to develop at grain boundaries rather than in the bulk material.

In stainless steels, most intergranular corrosion can be linked to metallurgical changes that occur within the temperature range of 427–816°C. These changes are often caused by slow cooling or prolonged welding processes. When stainless steels are exposed to these temperatures, they become sensitized to intergranular corrosion, as the formation of chromium carbides at the grain boundaries depletes the surrounding areas of chromium, reducing their corrosion resistance. This phenomenon is a critical concern in industries where welding or high-temperature processing is common (Figure 4.6).[4]

4.4.6 Selective Leaching

Certain metal alloys may experience a form of corrosion in which only one alloy element is removed and deteriorated selectively in specific corrosive environments. Selective leaching or dealloying is the term for this procedure. Dezincification, or the selective removal of zinc from brass alloys, is a typical example. This type of corrosion is especially dangerous for alloys made of metallic elements that are widely separated in the galvanic series. The various forms of selective leaching are frequently called after the alloy family that is impacted, usually according to the element that dissolves. A type of selective leaching known as dezincification involves the preferential removal of zinc from alloys containing zinc, especially brasses. This process mainly affects alloys with less than 85% copper. As a type of dealloying, dezincification was first observed in brass, where zinc dissolves and separates from copper. The term "dezincification" remains widely used. In brass, which typically consists of 30% zinc and 70% copper, this phenomenon results in a noticeable color change from yellow to red, making it easily identifiable by

FIGURE 4.6 Picture of intergranular corrosion.

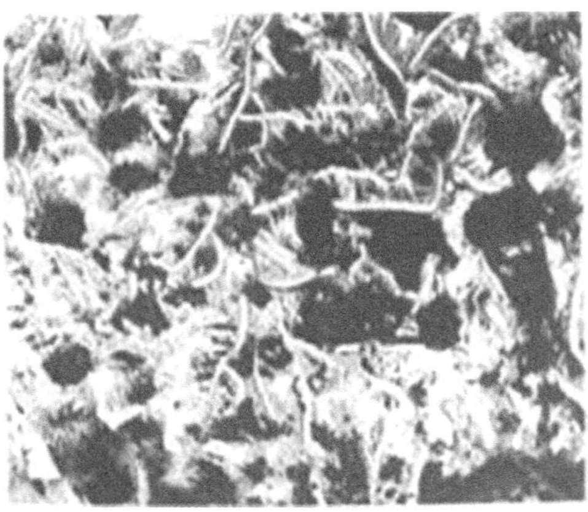

FIGURE 4.7 A pump impellor section showing graphite corrosion attack.

visual inspection. Dezincification usually occurs in stagnant water conditions. Copper-zinc alloys with more than 15% zinc are particularly susceptible to this type of corrosion. This phenomenon highlights the importance of material selection and environmental control in preventing selective leaching in industrial applications (Figure 4.7).

4.4.7 Stress Corrosion

When a metal fails as a result of both stress and chemical attack, stress corrosion takes place. This phenomenon, which results in component failure through the initiation and growth of high aspect ratio cracks, is caused by the interaction of static tensile stress, environmental factors, and, in certain cases, particular metallurgical conditions. Fine cracks that jeopardize structural integrity are its defining feature. This kind of failure, known as stress corrosion cracking (SCC), can happen suddenly and without warning, sometimes even after a few months or years of service that seemed to be dependable. SCC is a form of corrosion that occurs when tensile stress and a corrosive environment interact, leading to the brittle fracture of susceptible materials. While SCC may progress slowly, it can result in sudden and catastrophic failure, posing significant risks in industries such as aerospace, nuclear power, and chemical processing. Identifying the contributing factors, underlying mechanisms, and effective prevention strategies is essential for minimizing structural failures and ensuring the long-term reliability of critical components.

The following conditions are necessary for SCC to occur:

- A susceptible metal.
- A specific environment.
- A tensile or residual stress

The various forms of SCC are distinguished as follows:

(a) Chloride stress corrosion cracking. Austenitic steels undergo tensile stress in the presence of oxygen, chloride ions, and high temperatures.
(b) Caustic stress corrosion cracking. Steels can break in caustic environments with high hydrogen concentrations, such as Inconel tubes in alkaline solutions.
(c) Sulfide stress corrosion cracking. Steel cracking in hydrogen sulfide environments, commonly seen in the oil drilling sector.
(d) Seasonal cracking. The phrase has become obsolete. It was solely historically significant. This only applies to brass SCC in an ammonia atmosphere, although it can also occur in refrigeration plants that use ammonia as a refrigerant (Figure 4.8).[13]

4.4.8 Erosion Corrosion

Corrosion occurs when a metal surface interacts with corrosive fluids. This process is called as erosion. Erosion occurs when particles in a fluid are harder than the metal surface, resulting from corrosion and abrasion. When the particles are softer, erosion occurs through corrosion and attrition. The relative flow rate gives an indication of abrasion. Erosion occurs due to the simultaneous formation and removal of surface scale, with chipping and cracking mechanisms determining its cause. Cavitation corrosion is a specific type of erosion, occurring when vapor bubbles form and collapse near the metal surface in a liquid, leading to material degradation.

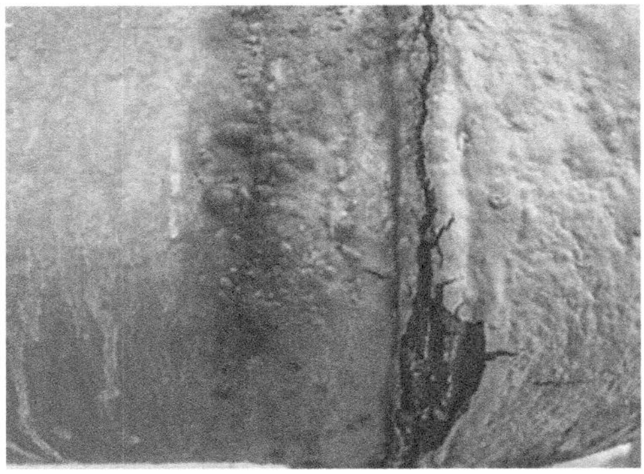

FIGURE 4.8 Image of stress corrosion.

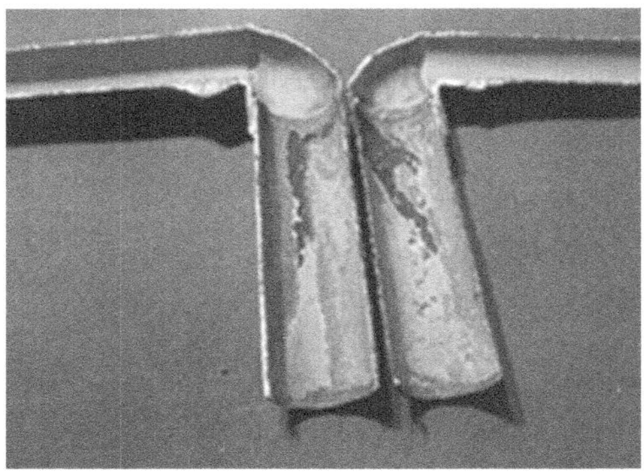

FIGURE 4.9 Erosion corrosion damage to copper tube.

Wear caused by solid particle erosion-corrosion has attracted significant research interest in recent years due to its detrimental effects on service components, leading to material loss and potential equipment failure. Erosion-corrosion is a common challenge in power plants, oil and gas processing facilities, and chemical plants, where solid particles, corrosive fluids, and target materials interact. This issue affects both static equipment—such as pipelines, valves, heat exchangers, and pressure vessels—and rotating equipment, including compressors, turbines, and pumps. Selecting suitable materials for these demanding environments is crucial, as the extreme conditions can greatly accelerate wear and degradation.[5]

The primary factors contributing to erosion-corrosion damage include water chemistry, temperature, and fluid velocity. Synergy plays a key role in this process, referring to the additional wear rate experienced by a metal under the combined effects of erosion and corrosion. This synergistic effect results in a wear rate that is greater than the sum of the wear rates caused by pure erosion and flow corrosion alone. Addressing erosion-corrosion requires careful consideration of material properties, environmental conditions, and operational parameters to minimize damage and extend equipment lifespan (Figure 4.9).

4.5 CORROSION PREVENTION METHODS

One of the biggest problems facing humanity is corrosion, which costs US$22 trillion a year, or 3% of the world's gross domestic product.[6] The goal of corrosion control measures is to completely eliminate corrosion in a limited number of situations. The field of corrosion prevention and protection techniques has seen substantial advancements over time. Noble metals don't corrode in general, but they are too expensive to be used for everyday applications.[1] As a result, we must fabricate various types of machinery using different metals and alloys to prevent corrosion. The three main subgroups of external corrosion protection and control techniques passive, active, and hybrid protection methods are based on information gathered from 122 pertinent studies.

4.5.1 Passive Corrosion Method

Passive corrosion protection functions by electrically insulating underground pipelines from external corrosive factors and preventing corrosion. This method encompasses coatings and linings, barrier applications or wrapping, strategic selection and design of pipe materials, electrical isolation, chemical inhibitors or treatments, and multi-passive protection strategies.

Coatings and Lining

Coatings are essential for preserving the integrity of buried pipes because they act as the first line of defense against corrosion. When used correctly, these protection techniques can protect more than 99% of the pipeline surface by acting as a physical barrier between the metal and its surroundings. However, a number of things can lead to coating defects, such as malfunctioning coating machines, incorrect handling, rock or debris penetration during backfilling, cracking from extreme mechanical stresses, soil microbiological activity, or damage from later construction activities.

Stray currents may start to flow when a coating defect arises, forming an anodic region at the defect's exit and a cathodic region at its entrance. This causes the pipeline's exposed surface to corrode more quickly. In these situations, the type of material and the surrounding environment can have a substantial impact on the corrosion rate. To avoid flaws and guarantee long-term corrosion protection, proper coating application, routine inspection, and maintenance are crucial.

4.5.2 Active Corrosion Method

Techniques for actively preventing corrosion on metallic surfaces are how active corrosion protection works. By applying a protective current, CP transforms all of the anodic (active) areas of a metal pipeline into cathodic (passive) areas. Humphry Davy used iron anodes to shield copper sheathing from corrosion caused by seawater in London in 1824, which is the earliest known application of CP. By the early 1940s, CP was implemented to safeguard gas steel pipelines and cast-iron water mains, significantly reducing leakage rates. The two most used CP techniques are galvanic or sacrificial anode cathodic protection and impressed current cathodic protection.[7]

4.5.3 Hybrid Corrosion Method

To create a more effective defense system against external corrosion in metallic pipelines, active and passive techniques are commonly combined. To reduce corrosion and improve protection against particular corrosion factors, this hybrid approach can be applied to existing pipelines or used during the initial pipeline installation. Defects, newly discovered corrosion causes, and coating, barrier, or lining degradation all affect how CP is integrated with passive techniques.

4.6 CORROSION TESTING METHODS FOR AMINO ACIDS

Numerous methods have been used to assess glutamic acid's effectiveness in preventing corrosion and to look at the protective layer that forms on metal surfaces. These techniques include measurements of hydrogen evolution, thermometric techniques, microscopy techniques, and electrochemical techniques like open-circuit potential, polarization curve analysis, and electrochemical impedance spectroscopy. Glutamic acid adheres to various adsorption isotherms and can act as a cathodic, anodic, or mixed-type inhibitor, depending on the metal type and the corrosive environment.[3] In recent years, computational chemistry techniques, such as semi-empirical methods and density functional theory, have been employed to investigate the impact of amino acid molecular structures on their corrosion inhibition efficiency. In addition, molecular dynamics simulations with periodic boundary conditions, incorporating the Metropolis Monte Carlo statistical method, have been utilized to compute adsorption energy and analyze the adsorption configurations of different amino acids on metal surfaces.

Corrosion testing for amino acids involves determining the propensity for materials to degrade when exposed to amino acid solutions. Under some situations, amino acids can be corrosive, particularly in industries that employ them frequently, such as pharmaceuticals, food processing, and biotechnology. Here is a full description of different corrosion testing methods:

Weight Loss Approach: In this approach, a clean metal specimen is exposed to an amino acid solution under regulated circumstances (temperature, concentration, pH) for a set amount of time. Following exposure, the specimen is removed, cleaned, and weighed to evaluate how much material has been lost due to corrosion. By comparing the starting and end weights, the corrosion rate can be determined.

Electrochemical Methods:

Potentiodynamic Polarization: This approach measures the polarization behavior of a metal electrode in an amino acid solution. By varying the electrode potential, the corrosion potential (E_{corr}) and corrosion current density (i_{corr}) can be calculated. These parameters provide insight into the material's corrosion resistance.

Electrochemical Impedance Spectroscopy determines the impedance of a metal-electrolyte interface as a function of frequency. Impedance spectrum analysis can provide information regarding corrosion processes such as corrosion rate and mechanism.

The Salt Spray Test (ASTM B117) is primarily used to test general corrosion resistance, but it can also be used to examine the corrosion resistance of materials in amino acid environments. In this test, metal specimens are exposed to a salt spray containing amino acids for a set amount of time. The appearance and severity of corrosion on the specimens are then assessed visually or using metrics such as mass loss.

Immersion Test: In this approach, metal specimens are immersed in amino acid solutions for an extended period of time under particular circumstances (such as temperature and pH). The specimens are evaluated on a regular basis for signs of corrosion, such as discoloration, pitting, or weight loss. The severity of corrosion is assessed using visual inspections or quantitative measurements.

Localized Corrosion Tests:

Pitting Corrosion Test: Pitting corrosion is a popular type of localized corrosion. Pitting resistance can be assessed by exposing metal specimens to amino acid solutions and monitoring pit start and propagation with techniques such as microscopy or electrochemical analysis.

Crevice Corrosion Test: Crevice corrosion occurs in small places like gaps, joints, or beneath deposits. Test specimens with simulated fissures are exposed to amino acid solutions, and the extent of crevice corrosion is determined visually or via metrics like depth of attack. These corrosion testing methods aid in determining the compatibility of materials with amino acid environments, as well as guiding material selection and design considerations to address corrosion-related difficulties in a variety of sectors.

4.7 GREEN CHEMISTRY

Green chemistry is an emerging field focused on promoting sustainability at the molecular level. Over the past decade, it has gained significant attention due to its potential to achieve both environmental and economic goals through innovative chemical design.[8] Green chemistry emphasizes careful planning of chemical synthesis and molecular design to minimize negative environmental impacts. By prioritizing effective design, it aims to create synergies rather than trade-offs, ensuring that chemical processes are both sustainable and economically viable. Numerous industries, including aerospace, automotive, electronics, cosmetics, energy, household goods, pharmaceuticals, and agriculture, have used this strategy. Numerous successful, award-winning technologies have demonstrated that green chemistry can deliver competitive and environmentally friendly solutions.

The potential of amino acids, which are physiologically significant molecules with both carboxylic acid (–COOH) and amine ($-NH_2$) functional groups, as corrosion inhibitors has been investigated. The ability of amino acids like alanine (Ala), methionine (Met), aspartate (Asp), asparagine (Asn), lysine (Lys), arginine (Arg), and histidine (His) to inhibit corrosion has been studied.[9]

One of the most practical and efficient ways to protect metals in corrosive environments is to use corrosion inhibitors. Organic compounds are thought to be very effective corrosion inhibitors, especially those that contain heteroatoms like nitrogen (N), sulfur (S), phosphorus (P), and oxygen (O), as well as numerous bonds and functional groups. By adhering to metal surfaces, these molecules create protective coatings that slow down the rate of corrosion.

Researchers hope to create sustainable and efficient corrosion protection solutions that meet industrial and environmental demands by fusing the ideas of green chemistry with the use of eco-friendly inhibitors like amino acids. In addition, it has been observed that organic molecules with the following configurations are effective corrosion inhibitors, especially in acidic environments: –OH, –COOH, NH_2, etc.[3] One example of an encourager material that is safe to use as a corrosion inhibitor is amino acids. They are biomolecules that are important to all organisms and serve as the building blocks of proteins as well as many other chemicals such as neurotransmitters, hormones, and nucleic acids. Amino acids are environmentally friendly beneficial molecules that are totally soluble in aqueous solutions, can be manufactured at a minimal cost, and are nontoxic. Those characteristics would support their usage as corrosion inhibitors. In 1906, 1907, and 1908, Hermann Leuchs et al.[1–3] published three articles on the synthesis and characteristics of α-amino acid N-carboxyanhydrides (NCAs, 1; Scheme 1), also known as Leuchs' anhydrides. Amino acids operate as corrosion inhibitors because of their ability to form protective coatings on metal surfaces. Here is an overview of how amino acids act in this context:

1. **Adsorption:** Amino acids can adsorb onto metal surfaces because they contain functional groups like amino (–NH_2) and carboxyl (–COOH). This adsorption process creates a protective coating that protects the metal from corrosive substances.
2. **Film Formation**: Once adsorbed on a metal surface, amino acids can undergo chemical reactions that result in a persistent film. This coating functions as a barrier, keeping corrosive species like oxygen and water from accessing the metal surface and causing corrosion.
3. **Passivation:** Amino acids may promote the production of passive oxide layers on metallic surfaces. These oxide layers are more stable and less prone to corrosion than the underlying metal, which increases the metal's corrosion resistance.
4. **Inhibition of Cathodic and Anodic Reactions:** Amino acids can also hinder the cathodic and anodic processes that occur throughout the corrosion process. By reducing these electrochemical reactions, amino acids slow corrosion and increase the life of metal components.
5. **Environmentally harmless:** Since amino acids are biodegradable and nontoxic, they are frequently seen as more environmentally benign than typical corrosion inhibitors. This makes them suitable for a variety of industrial applications in where environmental concerns are crucial.
6. **Synergistic Effects:** Amino acids can have synergistic effects when combined with other corrosion inhibitors or additives. This synergy can improve the overall corrosion inhibition performance and the efficiency of corrosion prevention techniques.

4.8 AMINO ACIDS CORROSION INHIBITION TYPES AND ENVIRONMENT

A chemical substance known as a corrosion inhibitor lowers the rate of corrosion to a manageable level when added to a corrosive environment.[10] Each amino acid has a side chain that is specific to it, in addition to amine (–NH_2) and carboxyl groups (–COOH).

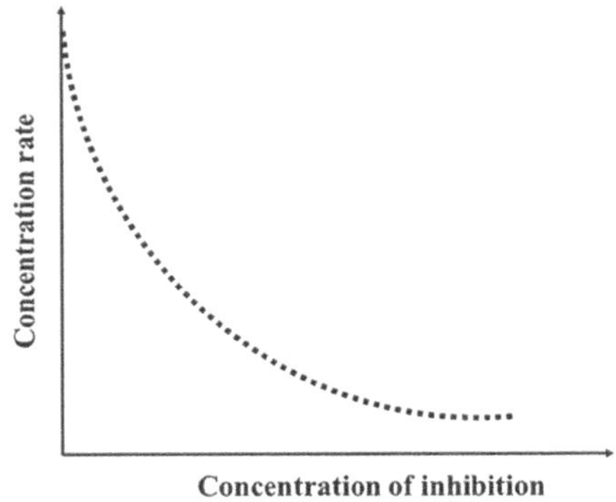

FIGURE 4.10 Corrosion rate as a function of inhibitor concentration.

The N and O atoms have a single pair of electrons, which makes it easier for them to form chemical bonds with the Cu atoms on their surface (Figure 4.10).

Figure 4.10 shows the relationship between the concentration of an inhibitor to the corrosion rate.

There are two ways that a corrosion inhibitor can aid in corrosion prevention. To make an environment less corrosive, corrosion inhibitors can interact with corrosive species. Corrosion inhibitors interact with metal surfaces to prevent corrosion. Inhibitors are classified into two types based on their manner of interaction: environment modifiers and adsorption inhibitors. Environment modifiers suppress corrosive species by interacting with them, preventing metal assault. This is a prime example where oxygen scavengers, such as hydrazine or sodium sulfite, along with cobaltous nitrate and biocides, can help mitigate microbiological corrosion. In neutral and alkaline solutions, oxygen reduction serves as a cathodic reaction that contributes to corrosion. However, the use of oxygen scavengers can suppress this reaction, thereby preventing corrosion.10 The field of corrosion inhibitors is currently undergoing significant changes, driven by the growing demand for environmentally friendly solutions. Regulatory agencies worldwide have imposed strict rules regarding the use and discharge of corrosion inhibitors, focusing on criteria such as toxicity, biodegradability, and bioaccumulation. As a result, there is a growing emphasis on developing eco-friendly or green corrosion inhibitors that have minimal or no negative environmental impact.

Eco-friendly corrosion inhibitors, like conventional ones, are categorized into organic and inorganic types. Research has highlighted the potential of naturally occurring substances, such as plant extracts, as effective green corrosion inhibitors for metallic materials in various corrosive environments. These natural compounds are not only environmentally benign but also cost-effective and renewable. A key factor driving interest in eco-friendly inhibitors is the presence of heteroatoms (e.g., sulfur (S), nitrogen (N), and oxygen (O)) and conjugated π-electron systems in their molecular structures. These features enhance their ability to adsorb onto metal surfaces, forming protective layers that inhibit corrosion. Scientists are increasingly exploring these properties to design and optimize green corrosion inhibitors that are both effective and sustainable. This shift toward environmentally compatible corrosion

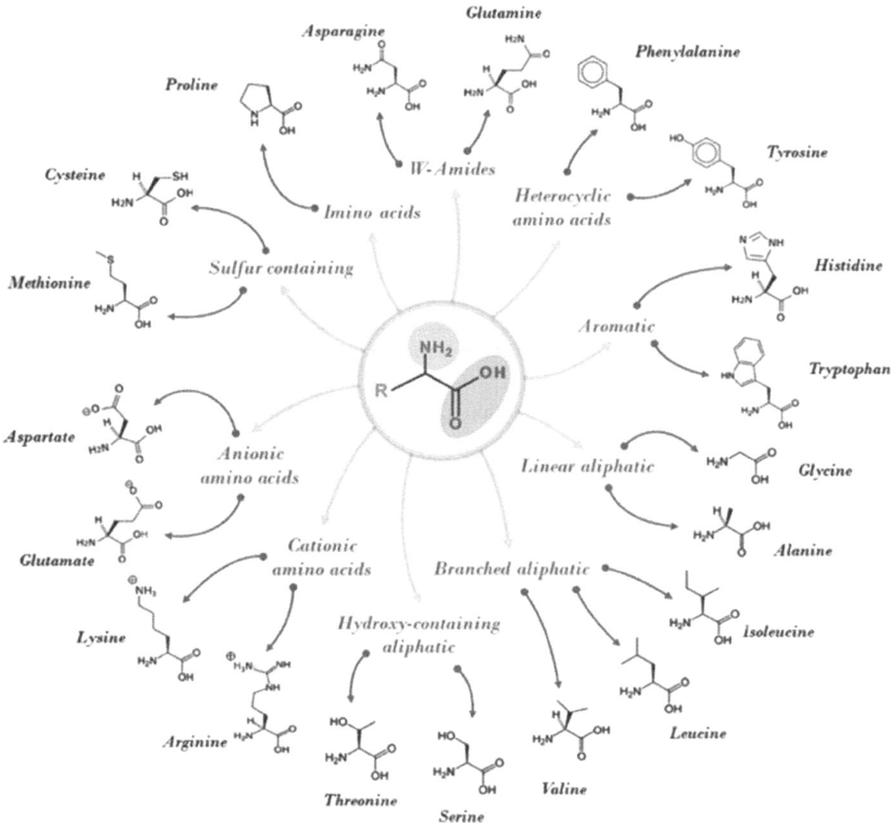

FIGURE 4.11 Structures of the 20 physiological amino acids and their chemical classification.[4]

inhibitors aligns with global efforts to reduce the ecological footprint of industrial processes while maintaining high performance in corrosion protection (Figure 4.11).[3]

4.9 COPPER AND ITS ALLOYS

Copper is a metal with a wide range of applications due to its excellent qualities, particularly its electrical and thermal properties. It is used in electronics to make wires, tubes, and sheets, as well as to build alloys from them. Copper is resistant to the effects of the atmosphere and numerous chemical species. Copper and its alloys are frequently utilized in electronics and integrated circuit manufacturing due to their high electrical and thermal conductivity. Copper, a noble metal, requires powerful oxidants for corrosion and disintegration. Chemical dissolution and electrolytic plating are the primary methods used. In the production of electronic devices, nitric acid is the most commonly used corrosive solution, leading to extensive research on copper corrosion. Organic inhibitors have a crucial role in protecting copper against corrosion. A review of the literature indicates that organic compounds and their derivatives, such as azoles, amines, and amino acids containing sulfur, have shown inhibitory effects on the corrosion of copper and its alloys in various environments. The presence of heteroatoms like nitrogen, sulfur, oxygen, and phosphorus in these organic molecules enhances their effectiveness as copper corrosion inhibitors. This is attributed to the vacant d orbitals in copper atoms, which allow the formation of coordinative bonds with electron-donating atoms.

However, many of these compounds are synthetic, often making them expensive, toxic, and hazardous to both living organisms and the environment. For example, benzotriazole and its derivatives are highly effective corrosion inhibitors for copper and its alloys but are known for their toxicity.

The corrosion inhibition of copper in 1 M HNO_3 was analyzed using gravimetric and potentiodynamic polarization methods with amino acids such as aspartic acid (Asp), glutamic acid (Glu), alanine (Ala), asparagine (Asn), glutamine (Gln), leucine (Leu), methionine (Met), and threonine (Thr). The results from both techniques were consistent, indicating that methionine (Met) exhibited the highest inhibition efficiency for copper in 1 M HNO_3. In addition, a correlation between inhibition efficiency and quantum chemical calculations was explored using semi-empirical (AM1 and MNDO) and ab initio methods (Table 4.1).[11]

4.10 STEEL AND ITS ALLOY

Steel is widely utilized due to its advantageous properties; however, its susceptibility to corrosion remains a significant drawback, which can be mitigated through coatings or protective paints.[3] The corrosion inhibition efficiency of amino acids—leucine, alanine, methionine, and glutamic acid—on NST-44 carbon steel has been evaluated using the weight loss immersion method and optical microscopy. Results indicate that alanine exhibited the highest inhibition efficiency in cassava fluid, averaging approximately 50%, while glutamic acid demonstrated the lowest efficiency, at less than 23%.[3]

Alagbe et al. further investigated the inhibitory effects of these amino acids on NST-44 mild steel in lime fluid (*Citrus aurantifolia*) using weight loss and optical microscopy techniques. Their findings reaffirmed alanine as the most effective corrosion inhibitor, whereas glutamic acid had minimal impact. Across both studies, glutamic acid consistently proved ineffective in preventing steel corrosion in cassava solution and lime fluid.

Electrochemical techniques have been employed to examine the corrosion behavior of steel in 0.5 M HCl in the presence of four amino acids: glycine, threonine, phenylalanine, and glutamic acid. The findings indicated that glutamic acid failed to establish a protective barrier against aggressive agents on steel surfaces. In amino acid solutions, a distinct anodic reaction rate was observed, accompanied by some resistance. Depending on the corrosive environment and the metal type, glutamic acid follows different adsorption isotherms and can act as a cathodic, anodic, or mixed-type inhibitor. To assess the corrosion inhibition efficiency of glutamic acid and characterize the protective film formed on metal surfaces, various techniques have been utilized, including weight loss analysis, electrochemical measurements (such as open-circuit potential, polarization curve method, and electrochemical

TABLE 4.1
Summary of experimental conditions, techniques, and inhibition efficiencies of various amino acids as corrosion inhibitors for copper in acidic media

Metal	Medium	Inhibitor	Experimental Techniques	Results
Copper	0.5 M HCL	Aspartic acid Glutamic acid Asparagine Glutamine	Potentiodynamic Polarization and electrochemical impedance spectroscopy	A 0.1 M concentration of glutamic acid demonstrated an inhibition efficiency of 59.70%. Aspartic acid, glutamic acid, asparagine, and glutamine function as mixed-type inhibitors. Their adsorption behavior follows the Langmuir adsorption isotherm, and the free energy of adsorption (ΔG_{ads}) indicates strong physical adsorption.
	1 M HNO_3	Aspartic acid Glutamic acid Alanine Asparagine Glutamine Leucine Methionine Threonine	Weight loss method, electrochemical method, and quantum chemical calculations	Methionine (80.38%), glutamic (37.55%) Good correlation between inhibition efficiency and quantum chemical parameters
	0.5 M HCL	Serine Threonine Glutamic acid	Electrochemical method and quantum chemical calculations	1 mM glutamic acid (94.5%)
	0.5 M HCL	Glycine Threonine Phenylalanine Glutamic acid	Electrochemical methods	In the presence of glycine, threonine, phenylalanine, and glutamic acid, the corrosion rate of copper was decreased glutamic acid (53.6%)
	0.5 M HCL	Glutamic acid Cysteine Glycine and their derivatives	Electrochemical impedance spectroscopy and cyclic voltammetry and quantum chemical calculations	Glutathione > cysteine > cysteine + glutamic acid + glycine > glutamic acid > glycine

impedance spectroscopy), hydrogen evolution measurements, thermometric techniques, and microscopy analyses. Despite its limitations, glutamic acid has shown potential as a corrosion inhibitor for metals in certain aggressive environments. A summary table provides reviewed examples, detailing the studied metal, corrosive solution, tested substances (glutamic acid and other amino acids), with or without additional compounds, and the key findings, including inhibition efficiency, based on experimental and computational methods (Table 4.2).

TABLE 4.2
Corrosion Inhibition performance of selected amino acids on steel corrosion in acidic media

Metal	Medium	Inhibitor	Experimental Techniques	Results
Steel	0.5 M HCL	Glycine Threonine Phenylalanine Glutamic acid	Electrochemical methods	Phenylalanine (74.8%) > glycine (53.9%) > Glutamic acid (8.2%)
NST-44 carbon steel	Cassava fluid: hydrocyanic acid	Leucine Alanine Methionine Glutamic ACID	Weight loss method and optical microscopic technique	Alanine (50%) > leucine (46%) > methionine (45%) > glutamic acid (30%). Alanine, leucine, and methionine characterized by equal amino group and carboxylic group while glutamic acid contains one amino group and tow carboxylic groups in the molecule in this case acidic solution attack the steel because glutamic acid cannot adsorbed on the surface of metal
NST-44 mild steel	Lime fluid	Leucine Alanine Methionine Glutamic acid	Weight loss method and optical microscopic technique	Methionine (44.70%) > Alanine (43.93%) Leucine (40.97%) > Glutamic acid (23.04%)
Carbon steel	Sea water	Glutamic acid With Zn^{2+}	Weight loss method, electrochemical method, and AFM analysis	200 ppm glutamic acid with 25 ppm Zn^{2+} (EI = 87%). The inhibitor (glutamic acid–Zn^{2+}) binds to surface of carbon steel

4.11 IRON AND ITS ALLOYS

Amino acids contribute significantly to corrosion reduction by facilitating the formation of a passive layer on the surface of iron and its alloys. This protective barrier prevents corrosive agents from penetrating the metal surface. Through coordination with metal ions, amino acids form stable complexes that enhance the stability and integrity of the passive film. By promoting passivation, amino acids effectively reduce corrosion rates and extend the service life of iron-based materials.

The ability of amino acids to inhibit corrosion is primarily attributed to their capacity to form stable complexes with metal ions. Functional groups such as $-NH_3$ and –COOH facilitate interactions with metal surfaces, leading to the formation of chelate complexes. These complexes decrease the availability of metal ions for corrosion reactions, thereby slowing the corrosion process. In addition, the formation of such complexes enhances amino acid adsorption onto the metal surface, further strengthening the protective barrier against corrosion.

Amino acids inhibit corrosion through adsorption onto the metal surface, forming a protective layer. This adsorption is facilitated by various interactions, including:

- Hydrogen bonding: between amino acid functional groups and the metal surface.
- Electrostatic interactions: between charged groups on the amino acid and the metal surface.
- Coordination bonding: between amino acid functional groups and metal ions.

Amino acids, once adsorbed onto metal surfaces, form a physical barrier that blocks corrosive species from reaching the metal. In addition, they can undergo chemical interactions with the metal surface, leading to the formation of additional protective layers that further inhibit corrosion. Some amino acids also exhibit pH buffering capabilities, which play a role in controlling the corrosion behavior of iron and its alloys. By maintaining an optimal pH and stabilizing the passive layer on the metal surface, amino acids help reduce the formation of aggressive corrosion products. This pH buffering effect contributes to lowering corrosion rates and extending the service life of iron-based materials in harsh environments.

Research by Hluchan et al. investigated the corrosion inhibition effects of 22 amino acids and four related compounds on iron in a deaerated 1 M HCl solution. The study found that while some compounds, such as hydroxyproline, creatine, cystine, and cysteine, acted as corrosion accelerators, others showed significant inhibitory effects. Tryptophan emerged as the most effective amino acid, and 3,5-diiodotyrosine was the most effective related compound, with inhibition efficiencies of 80% and 87% at 10 mM, respectively. This high efficiency was attributed to the presence of favorable adsorption sites, such as aromatic rings and iodine atoms, in their molecular structures.

In another study, Zerfaoui et al. examined the effects of five amino acids (glycine, leucine, aspartic acid, arginine, and methionine) in citric acid media using weight loss measurements, DC, and AC electrochemical techniques. They found that methionine was the most effective inhibitor, with its performance strongly influenced by the solution's pH. Methionine's inhibition efficiency increased with pH, reaching up to 90% at pH 5.

These studies highlight the potential of amino acids as eco-friendly corrosion inhibitors, with their effectiveness depending on molecular structure, adsorption properties, and environmental conditions like pH. Their ability to form protective barriers, stabilize passive layers, and buffer pH makes them promising candidates for sustainable corrosion protection in various industrial applications. In addition, polarization measurements indicated that the tested substances function as cathodic inhibitors. Ashassi-Sorkhabi et al. examined the impact of hydrodynamic conditions on the inhibition efficiency of methionine against mild steel corrosion in a molar sulfuric acid solution. A key finding was that in the absence of an inhibitor, the metal surface potential becomes more noble as the rotation speed increases. The inhibitory efficiency for still electrodes and low-speed rotations did not change substantially; however, at higher rotation speeds the efficiency increased. The correlation between the inhibition efficiency of amino acids and their chemical structure has been studied for the corrosion of Armco iron in 1 M HCl at 308 K. Weight loss measurements were conducted using Armco iron specimens immersed in 1 M HCl solutions, both with and without the addition of various amino acids, including methionine, cysteine, cystine, glycine, leucine, arginine, serine, glutamic acid, ornithine, lysine, aspartic acid, alanine, valine, asparagine, glutamine, and threonine.

The study revealed that the most effective inhibitors were methionine, cysteine, and cystine, all of which contain a sulfur atom in their molecular structure. These amino acids adsorb onto the metal surface primarily through their sulfur active centers. The extent of inhibition is directly related to the formation of an adsorption layer, which is highly dependent on the molecular structure of the inhibitor.

For methionine, the corrosion inhibition mechanism is explained by the Frumkin isotherm model, which describes the adsorption of inhibitor species on the Armco iron surface. The inhibition efficiency is related to the fraction of the electrode surface blocked by the adsorbed inhibitor molecules.

In another experiment, the effect of amino acids on the corrosion of iron in 1 M HCl was evaluated using weight-loss measurements after a 6-h immersion period at 308 K. The results confirmed that sulfur-containing amino acids, such as methionine, cysteine, and cystine, exhibited the highest inhibition efficiencies due to their strong adsorption on the metal surface.

These findings emphasize the crucial role of molecular structure, especially the presence of sulfur atoms, in influencing the corrosion inhibition efficiency of amino acids.[12] The capacity of these compounds to form stable adsorption layers on the metal surface significantly enhances their protective performance, positioning them as promising candidates for environmentally friendly corrosion prevention in acidic environments.[12] Inhibition efficiency ($E_w\%$) is calculated as follows:

$$E_{NK}\% = \left(1 - \frac{w}{w^{\circ}}\right) \times 100.$$

where w and w° are the corrosion rates of iron samples in the absence and presence of the organic compounds, respectively.

The cathodic and anodic polarization curves of iron in molar HCl, both in the absence and presence of these molecules at a concentration of 10^{-3} M, are shown in Figures 4.12 and 4.13, respectively. The results indicate that all tested amino acids at

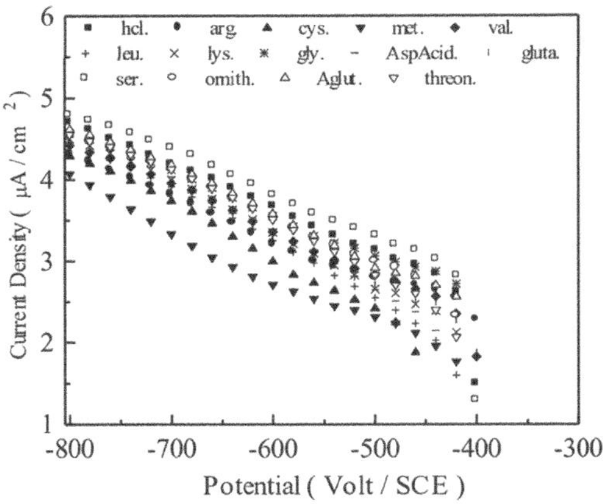

FIGURE 4.12 Cathodic potentiokinetic polarization curves of Armco iron in 1 m HCL in the presence of different amino acids.

Source: Aouniti et al. (2013).

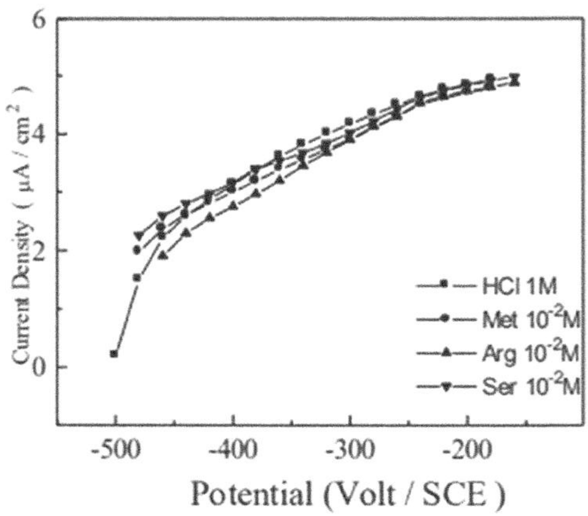

FIGURE 4.13 Anodic potentiokinetic polarization curves of Armco iron in 1 M HCL in the presence of some amino acids.[5]

10^{-3} M inhibit only the cathodic process. Among them, methionine (Met), cysteine (Cys), and cystine (Cyt), which contain sulfur in their molecular structures, were found to be the most effective inhibitors.

The studied organic compounds exert their inhibitory effect by slowing down the cathodic hydrogen evolution reaction. In the polarization method, no Tafel linearity was observed, particularly with Met, Cys, Cyt, and Arg. Consequently, the inhibition efficiency ($E_{pot}\%$) is determined using a specific relation: $E_{pato}\% = \left(1 - \frac{I_{corr}}{I^{\circ}_{corr}}\right) \times 100.$

where $I_{coor} = I_{coor}/A$, A is the exposed surface area and I_{coor} and I°_{coor} are the uninhibited and inhibited corrosion current densities, respectively, for a given ($E_{pot}\%$)

The effectiveness of compounds in preventing adsorption is influenced by various aspects such as the number of active centers, charge density, size, mode of adsorption, and formation of metallic copolymers. Corrosion inhibition relies on the presence of electron donors (N, O, S) in the amino acid structure. The presence of free electron pairs in nitrogen and sulfur atoms, as well as an electron on a double bond, promotes inhibitor adsorption.

REFERENCES

1. Ibrahimi B El, Nardeli JV, Guo L. An overview of corrosion. *ACS Symp Series*. 2021;1403:1–19. doi:10.1021/bk-2021-1403.ch001
2. Forbes K. 1993. *In theory and practice*. [place unknown]. doi:10.1097/00002800-199301000-00006
3. El Ibrahimi B, Jmiai A, Bazzi L, El Issami S. Amino acids and their derivatives as corrosion inhibitors for metals and alloys. *Arabian J Chem*. 2020;13(1):740–771. doi:https://doi.org/10.1016/j.arabjc.2017.07.013
4. Kolesar SC. texts't-'l.
5. Rajahram SS, Harvey TJ, Wood RJK. Erosion–corrosion resistance of engineering materials in various test conditions. *Wear*. 2009;267(1):244–254. doi:https://doi.org/10.1016/j.wear.2009.01.052
6. Zehra S, Mobin M, Aslam R. CHAPTER 2- Corrosion prevention and protection methods. In: Guo L, Verma C, Zhang D, editors. *Eco-friendly corrosion inhibitors*. [place unknown]: Elsevier; 2022. pp. 13–26. doi:https://doi.org/10.1016/B978-0-323-91176-4.00023-4
7. Farh HMH, Ben Seghier MEA, Zayed T. A comprehensive review of corrosion protection and control techniques for metallic pipelines. *Eng Failure Analysis*. 2023; 143:106885. doi:https://doi.org/10.1016/j.engfailanal.2022.106885
8. Anastas P, Eghbali N. Green chemistry: Principles and practice. *Chem Society Reviews*. 2010;39(1):301–312. doi:10.1039/b918763b
9. Kaya S, Tüzün B, Kaya C, Obot IB. Determination of corrosion inhibition effects of amino acids: Quantum chemical and molecular dynamic simulation study. *J Taiwan Inst of Chem Eng*. 2016;58:528–535. doi:https://doi.org/10.1016/j.jtice.2015.06.009
10. Ituen E, Akaranta O, James A. Evaluation of performance of corrosion inhibitors using adsorption isotherm models : An overview. *American Chem Sci J*. 2017;18(1):1–34. doi: 10.9734/csji/2017/28976

11. Barouni K, Kassale A, Albourine A, Jbara O, Hammouti B, Bazzi L. Amino acids as corrosion inhibitors for copper in nitric acid medium: Experimental and theoretical study. *J of Materials and Environ Sci.* 2014;5(2):456–463.
12. Aouniti A, Khaled KF, Hammouti B. 2013. Correlation between inhibition efficiency and chemical structure of some amino acids on the corrosion of Armco iron in molar HCl. *Int J Electrochem Sci.* 8(4):5925–5943. doi:https://doi.org/10.1016/S1452-3981(23)14731-6
13. Ahmad, Z. *Principles of Corrosion and Corrosion*; Great Britain; Butterworth-Heinemann; 2006

5 Amino Acids for Drilling Fluid Systems

Kwamena Ato Quainoo, Cornelius Borecho Bavoh and Eric Broni-Bediako

5.1 INTRODUCTION

The effective management of reactive geological formations with sustainable and environment-friendly drilling fluids has emerged a growing demand. Swelling and migrating clay minerals, found abundantly in geological formations like mud-making shale, caving shale, heaving shale, and sloughing shale, pose both advantages and challenges in drilling operations. While they contribute to the structural integrity of formations by acting as effective binding agents with sedimentary matrices, their swelling propensity upon exposure to drilling fluids poses a significant threat to wellbore stability, leading to costly delays, equipment failures, and environmental hazards such as fluid invasion and wellbore collapse. The use of drilling fluids that leaves little or no impact on the reacting phase of the formation is the go-to option. Various factors including cost, environmental impact, regulatory adherence, and fluid compatibility influence the decision on drilling fluid selection. Among the most efficient clay stabilizers are oil-based drilling fluids (OBM). Oil does not engage with clay minerals, thereby ensuring nearly complete preservation of the formation's structural integrity when used as drilling fluids. Nevertheless, the use of OBM faces challenges such as increased expenses, potential environmental implications, handling, and compliance restrictions. A friendlier and more cost-effective alternative is the utilization of water-based drilling fluids. Nonetheless, the high susceptibility of water to reactive formations presents the challenge of clay swelling during drilling operations in such formations. Swollen clay weakens the formation exposing it to borehole integrity issues.

Conventional methods of addressing clay swelling involve designing drilling fluids crucial for maintaining wellbore stability and facilitating efficient drilling processes. The success of fluid designs for drilling largely depends on understanding the physical and mineralogical characteristics of the formation and its response to contact with drilling mud.[1] Commonly used effective clay stabilizers in drilling fluids include salt-based fluids, potassium chloride (KCl), and ammonium chloride. Potassium ions are particularly effective as they fit snugly into the spaces between

DOI: 10.1201/9781003502388-5

clay mineral layers, while the anionic component helps counteract osmotic pressure during interactions with clay-based fluids. However, KCl is most effective in higher concentrations, typically ranging from 3 to 37 wt% on the clay mineral surface. The use of high concentrations of KCl presents challenges in terms of mixing, handling, and disposal, especially during extensive drilling operations. In addition, the anionic component of KCl is considered toxic and can contaminate underground aquifers as pointed out by Zhou and Nasr-El-Din.[2] Research has shown that salt solutions containing 2% or more KCl have chloride levels exceeding 9,500 ppm, which is exceptionally high. Such elevated chloride levels can be detrimental to aquatic ecosystems, particularly harming freshwater organisms sensitive to changes in water chemistry. High chloride concentrations disrupt aquatic habitats and can adversely affect the health of fish and other aquatic life. Moreover, current government regulations limit the surface disposal of substances containing more than 3,000 ppm chloride, applicable to both on and off-lease operations. Consequently, there is a significant interest in the industry for developing low-cost, eco-friendly clay stabilizers with enhanced performance.[3–8]

Amidst this backdrop, the pursuit of green clay stabilizers emerges as a promising avenue toward aligning drilling operations with environmental friendliness. Green inhibitors, characterized by their environmentally benign nature and sustainable attributes, offer a compelling solution to the dual challenge of clay stabilization and environmental preservation. By harnessing the principles of green chemistry and eco-friendly formulations, these inhibitors aim to mitigate clay swelling while minimizing adverse ecological footprints associated with drilling activities.

This book endeavors to explore the evolving landscape of green inhibitors for clay stabilization in drilling fluids, with a focus on their formulation, mechanisms of action, efficacy, and environmental implications. Through a comprehensive review of existing literature, experimental findings, and case studies, this study aims to shed light on the potential of amino acids as effective green inhibitors to revolutionize drilling practices, enhance operational efficiency, and foster environmental sustainability in the energy sector. By elucidating the role of amino acids in mitigating clay swelling while minimizing environmental impact, this book seeks to contribute to the ongoing discourse on sustainable drilling practices and pave the way for the adoption of eco-friendly solutions in the exploration and production of hydrocarbons.

5.2 CLAY SWELLING AND ITS IMPACT DURING DRILLING

The global transition to alternative sources of energy (shale and other sensitive reservoir types) from conventional sources is driven by the decline rate of the conventional resources.[9] One common but sensitive challenge that this transition faces primarily stems from the presence of reactive minerals common in these sensitive formations. Shale formations such as the Barnett, Haynesville, Marcellus, Utica, and Eagle Ford have substantially impacted North America's energy reserves. These formations have transformed the United States into a major exporter of natural gas. Shale, however, is primarily made up of small particles of clay minerals and quartz, which are the main constituents of mud.[9] Some of these minerals, when exposed to water, can swell up to ten times its volume. Swelling of clay during drilling operations significantly impacts the drilling process and is a major factor in reducing permeability in hydrocarbon reservoirs.[10] This presents a challenge that can affect both the efficiency and safety of drilling operations including a possible damage to the formation itself.[11] The petroleum industry currently spends millions of dollars annually to address these effects and mitigate their impact on drilling operations. To stabilize a permeability enhanced formation, maintaining swelling inhibition interactions between drilling fluids and the reactive or sensitive formations is very important. To achieve this, a more effective drilling fluid is needed.

Presently, the standards for a suitable drilling fluid mandate the use of materials (additives) that are nontoxic, biodegradable, cost effective, and environmentally safe, while still preserving the primary roles of drilling fluid. These roles include: (1) removing cuttings from the borehole and facilitating their separation at the surface; (2) maintaining wellbore stability; and (3) avoiding damage to the production formation and others.[12] The commonest base fluid employed in drilling is water, although air and oil are sometimes used. This is because of the cost benefits of utilizing water as the base fluid for drilling. The problem here is that the water-based drilling fluid often show enhanced geochemical reactions between the drilling fluid and the clay minerals present in shale and other sensitive formations. These reactions can cause the clay layers to absorb many molecules of water that weakens and disperses the clay minerals, resulting in the eventual weakening of the formation. Civan[13] hinted that clay swelling is one of the primary reasons for formation damage. Over the years, the industry has employed both mechanical and chemical technologies to minimize the impact of reactive clays. The inclusion of chemical additives into the water-based fluid is a preventive technique considered to be cost effective for addressing clay swelling-related problems. The inclusion of salts such as KCl, NH_4Cl, and $CaCl_2$ has proven to be very effective. These salts are very effective; however, their potential is largely contingent on higher concentrations in the base fluid, typically ranging from 2 to 35 wt%. These high levels of salts in the base fluid can have adverse effects on the marine environment including affecting the potential of the drilling fluid's carrying capacity. Bruton and McLaurine[14] hinted that drilling fluids treated with salts are constrained for use because of concerns such as toxicity, density, corrosion, rheology, and fluid loss. In addition, laboratory analysis has shown that these salts can disintegrate clay formations which can further weaken the formation.[3,15] A search for green, effective, and cost-friendly chemicals is currently ongoing. Figure 5.1 shows a list of clay swelling inhibitors utilized in drilling fluids in open literature.

5.2.1 Evaluating Natural Amino Acids as Effective Bio-Inhibitors for Drilling Fluid Technology

Several studies investigating novel, sustainable, cost-effective, and potential green additives for clay stabilization in drilling fluids are currently ongoing. Recently, clay stabilizers derived from natural resources, often termed bio-inhibitors, have gained increasing attention in research circles over traditional inhibitors. This interest is driven by their excellent swelling inhibition capabilities, biodegradability, and ready availability, as noted by Rana et al.[16] and Quainoo et al.[17]. Their effectiveness in controlling swelling and their broader impacts are linked to their unique functional

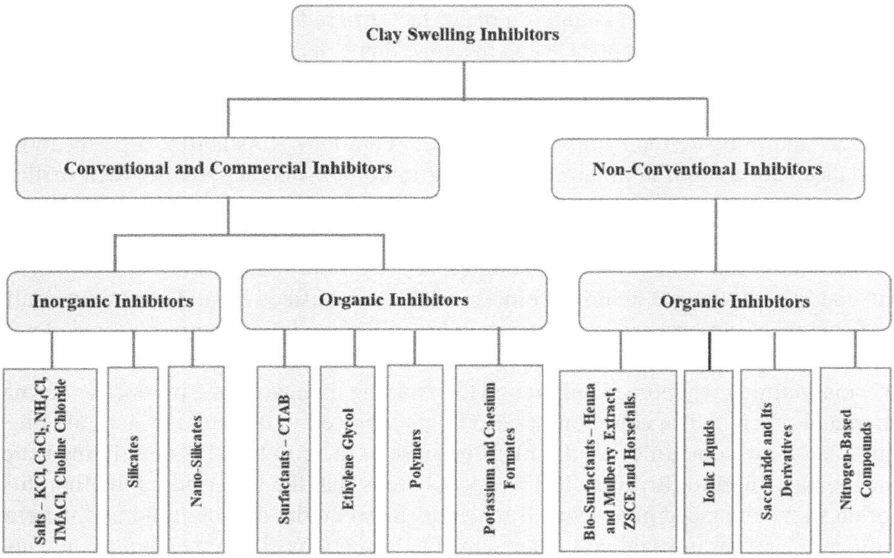

FIGURE 5.1 Sample list of clay swelling inhibitors utilized in water-based drilling fluids.

TABLE 5.1
Bio-Clay Stabilizers for Swelling Clays Studied in Open Literature

Name of Clay Stabilizer	Class of Bio-Inhibitor	Remarks	Reference
Methyl Glucoside	Saccharides	"Good inhibitor but works effectively under high concentration"	[18]
Chitosan Quaternary Ammonium Salt	Saccharides	"Good clay stabilizer"	[19]
Chitosan Grafted l-Arginine	Saccharides	"Showed better inhibition than KCl and displayed higher temperature resistance and longer durability"	[20]
Cationic Starch Derivatives	Saccharides	"Good potential shale inhibitor"	[21]
MLE	Biosurfactants	"Similar inhibitive performance as KCl and PA"	[22]
Horsetail Extract	Biosurfactants	"Performed worse than KCl and polyamine"	[23]
Ginseng	Biosurfactants	"Performance increases with increasing concentration"	[24]

(Continued)

TABLE 5.1
Continued

Name of Clay Stabilizer	Class of Bio-Inhibitor	Remarks	Reference
GGRE	Biosurfactants	"Better inhibitor than KCl"	[25]
Chromolaena odorata	Biosurfactants	"Comparable inhibition characteristics to KCl"	[26]
SRLSE	Biosurfactants	"Better inhibitor than KCl"	[27]
Pomelo Peel Extract	Biosurfactants	"Moderate shale inhibitor"	[28]
Tallow Amine Ethoxylate	Biosurfactants	"Comparable inhibition characteristics to KCl"	[29]
Tribulus terretris Extract	Biosurfactants	"Superior inhibition qualities to KCl and polyamine"	[30]
Henna Extract	Biosurfactants	"Deflocculating property, anti-corrosive/ moderate shale inhibitor"	[31]
ZSCE	Biosurfactants	"Has a high concentration of saponins that contributes to its performance"	[32]
Gelatin	Nitrogenous compounds	"Exhibits excellent inhibition properties with poor temperature stability"	[3]
Polyethyleneimine	Nitrogenous compounds	"Excellent clay stabilizer that shows inhibition performance better than HTCC and PEA"	[34]
OPLL	Nitrogenous compounds	"A polyelectrolyte with a strong inhibiting quality	[35]
Cycloaliphatic Amines	Nitrogenous compounds	Good inhibitor and highly temperature stable"	[36]

(Continued)

TABLE 5.1
Continued

Name of Clay Stabilizer	Class of Bio-Inhibitor	Remarks	Reference
Dopamine	Nitrogenous compounds	"Good clay stabilizer and is highly temperature resistant"	[37]
Poly(oxypropylene)-amidoamine	Nitrogenous compounds	"Good clay stabilizer. Hydrogen bonds are formed by the amide and siloxane groups"	[38]
Gelatin Quaternary Ammonium Salts	Nitrogenous compounds	"Excellent stability with enhanced temperature stability"	[39]
BHMT	Nitrogenous compounds	"A good inhibitor that is very resistant to high temperature conditions"	[40]

properties and they are believed to address shale destabilization through various inhibition mechanisms. Table 5.1 lists some bio-inhibitors tested for clay stabilization.

Although these bio-inhibitors were proven to be laboratory effective, more field evaluations are needed to confirm their effectiveness. Khodja et al.[41] indicated that some bio-inhibiting saccharides are susceptible to biological attack thus has a high degree of spoilage when stored on site. Some of these bio-inhibitors are also polymeric with large molecular weight that are capable of coating shales further limiting the permeability of the formation. A search for monomeric bio-inhibitors was therefore needed. The use of amino acids for stabilizing reactive clay formations was first evaluated by Quainoo et al.[43] The evidence for the first trial was based on prior work by several authors who investigated the clay stabilizing potentials of several oligomers of amino acids and other derivatives of amino acids. Chemicals such as oligo(poly-L-lysine, chitosan grafted l-arginine, poly arginine, HS, and many others were proven to be very effective.[14,35,45] For example, Bruton and McLaurine[14] of MI Drilling Fluids Co. evaluated the clay stabilizing potential of a modified poly amino acid in controlling reactive shales. They confirmed that amino acids are capable of suppressing the swelling of clays in both saline and nonsaline aqueous solutions and that their mud systems can be formulated without any adverse effects on the properties of the mud. Furthermore, amino acids were screen selected evaluated via analytical methods to evaluate their potential and they displayed positive results. It was in this light that several natural amino acids were then evaluated for stabilizing reactive clay formations.[43,42,44]

Natural amino acids are organic compounds typical identified by the formula, H2NCHRCOOH, and widely occur in nature. Amino acids, the building blocks of proteins and essential components of biological systems, possess unique properties that make them well-suited for interactions with clay minerals. They are composed of amine and carboxylic functional groups with distinct sidechains, which define their physical and chemical characteristics. In solution, the amine groups protonate to form ammonium cations. There are 22 naturally occurring amino acids with unique characteristics that can be evaluated for their clay stabilizing potentials. Their amphiphilic nature allows them to adsorb onto clay surfaces, forming protective layers that inhibit water penetration and reduce swelling tendencies. Furthermore, amino acids exhibit chelating abilities, enabling them to sequester metal ions and mitigate the adverse effects of drilling fluid contaminants on clay stability. So far over six natural amino acids and several derivatives of amino acids have been evaluated as potential drilling fluid clay stabilizing additives. Incredible results surpassing that of KCl were observed for some of the studied amino acids. In addition to their unique and effective functional properties, amino acids are also less expensive, and environmentally friendly. Regarding the need to impact the environment positively while ensuring the long-term sustainability of the reservoir formation, considering amino acids and their derivatives offer a greener alternative with potential benefits for both operational efficiency and environmental stewardship. Herein, we discuss the clay stabilizing potentials of amino acids and their derivatives

In recent years, research efforts have focused on elucidating the clay stabilizing potentials of amino acids, the mechanisms by which amino acids interact with clay minerals, their compatibility in drilling fluids, and their impact on drilling fluid's rheology[46]. One of the key advantages of amino acids as clay stabilizers lies in their eco-friendly profile. Unlike traditional inhibitors that may contain toxic or environmentally harmful components, amino acids are biodegradable and pose minimal risks to human health and the environment. This aspect aligns with the growing emphasis on sustainability and corporate responsibility within the energy industry, making amino acids an attractive choice for companies seeking to adopt greener drilling practices. This chapter provides amino acid-based clay stabilization technologies for drilling fluids by drawing upon a synthesis of experimental findings, field applications, and theoretical insights. This study will explore the mechanisms of action, efficacy, and environmental implications of amino acids as clay stabilizers. By evaluating the potential of amino acids in drilling practices and promote sustainability in the energy sector, this paper seeks to contribute to the ongoing discourse on environmentally responsible drilling operations. In essence, the utilization of amino acids as green clay stabilizers represents a promising avenue for enhancing wellbore stability while minimizing environmental impact. Through further research and innovation, amino acids have the potential to play a pivotal role in advancing sustainable drilling practices and shaping the future of the oil and gas industry.

5.3 THE MECHANISM OF AMINO ACIDS FOR STABILIZING REACTIVE CLAY FORMATIONS

Low and Margheim[47] suggested that the swelling pressure in a clay system, when exposed to water, is predominantly influenced by two key factors: (a) the concentration and type of ions present on the interlayer surfaces of the clay mineral, and (b) the surface area and reactive forces of the clays. The swelling of reactive clays typically occurs through two principal processes: crystalline and osmotic swelling. Crystalline swelling involves the restricted adsorption of water into the interlayer spacings of typical swelling clays, such as sodium or calcium montmorillonite. The structure of these clay minerals consists of negatively charged platelets, with the exhibition of these negative charges playing a crucial role in the clay's water sensitivity. To maintain electrical neutrality within the mineral, ions such as sodium and calcium serve as counterions. According to Segad et al.,[48] Monte Carlo simulations indicate significant swelling in clays with monovalent counterions, whereas clays with divalent counterions experience more restricted swelling, characterized by an aqueous layer of about 10 Å between the platelets. Consequently, it is suggested that calcium montmorillonite clay, which contains divalent ions, is exposed to water, it would exhibit less swelling compared to sodium montmorillonite clay, which contains monovalent ions. The counterions, such as calcium and sodium, play a crucial role in binding the clay platelets tightly together, thereby strengthening the ionic bridges between the interlayers. This binding restricts the entry of water into the interlayer spaces. On the other hand, osmotic swelling refers to the unlimited absorption of water due to increased water activity on the clay's surface. This type of swelling occurs when there is a discrepancy in ion concentration between the surface of the clay and the surrounding drilling fluid. This discrepancy creates an osmotic pressure gradient that drives water molecules into the clay interlayers, leading to expansion.[49] An effective clay stabilizer should effectively control both crystalline and osmotic swelling. Typically, the action of clay stabilizers in reactive clay minerals involves a dual or triple inhibition strategy, tailored to reduce swelling, and maintain structural stability during drilling operations. The more inhibitive mechanism a clay stabilizer possesses, the higher its stabilizing potential. The three primary mechanisms of inhibition include:

- *Counterion Replacement:* This mechanism involves replacing the natural counterions in the clay's interlayer spaces with robust cations that induce less swelling than sodium or calcium, effectively tightening the spacing between clay platelets. Aften and Ma[50] hinted that the intercrystalline space occupied by the sodium and calcium molecules in bentonite is gapped at 8–10 20 Å. These sodium and calcium molecules do not have sufficient size to hold tight the interlayer spaces thus becoming easily hydrated in the presence of water activity. A potent clay stabilizer has sufficient ionic size to hold the interlayer space together including reducing the hydration possibilities of these spaces in the presence of water activities. So far, potassium and ammonium cations have proven to have sufficient size to keep away many units of water from invading that space.
- *Reduction of Water Interaction:* This strategy reduces the interaction between the clay minerals and the drilling base fluid. It is often achieved using chemicals that strongly attract the base fluid, thereby limiting its interaction with the clay's surface. The figure below shows the many reactive sites of a given swelling clay. When the water activity between the clay's

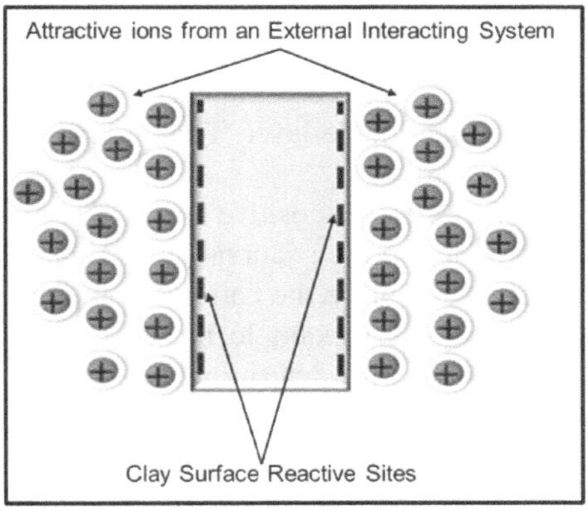

FIGURE 5.2 Schematic diagram showing the attraction of cations to a clay anionic surface.

surface and the base fluid is limited, there is an enhanced reduction in the interaction between the numerous reactive sites of the clay and the base fluid (see Figure 5.2).

- *Formation of Hydrophobic Shields:* Some clay stabilizers can create protective barriers around the clay particles using their hydrophobic components. These shields protect the clay from water molecules attempting to penetrate it. Chemicals that display this third inhibitive mechanism are mostly polymers.

Due to the distinct properties of each natural amino acid, they exhibit unique inhibitory mechanisms. However, the amino acids typically studied for clay stabilization processes have demonstrated two main inhibitive mechanisms: (1) They significantly disrupt the interaction between the water component of the drilling fluid and the reactive sites of the clay. Amino acids can form strong hydrogen bonds with the water in drilling fluids, thereby reducing extensive water penetration into clay formations and consequently, limiting clay swelling. Quainoo et al.[42] used the COSMO-RS tool to analyze the chemical interactions between six amino acids and water. This theoretical tool provided two critical thermodynamic parameters: the activity coefficient and hydrogen bonding energy. The activity coefficient indicates the likelihood of molecules within a solution escaping. A coefficient less than 1 suggests that attractive forces predominate in the mixture, requiring more energy to separate these molecules. Conversely, a coefficient greater than 1 indicates that repulsive forces are dominant, making it easier for molecules to separate. The hydrogen bonding energy (EHB) measures the electrostatic force of attraction between two molecules with strong dipole moments. Due to water's inherent polarity, it is capable of forming hydrogen bonds with polar amino acids. In addition, amino acids can help neutralize the numerous negative charges on the surface of clay. During drilling processes, when clay stabilizers are present, the positively charged cations of the stabilizers engage with the negatively charged surfaces of the clay, effectively neutralizing these charges. This neutralization reduces the clay's affinity for electrons, subsequently diminishing its sensitivity to water molecules. This action significantly curtails the infiltration of water into the swellable components of the formation, thus reducing the clay's micelle and electrical double layer. This reduction plays a crucial role in protecting the geological formation.

(2) Amino acids are rich with amine positive ions. These amine positive ions often protonate in solution to form ammonium cations (NH^{4+}). These cations upon interaction with the clay mineral can displace the cationic component present in the interlayer sites of a given clay mineral. For example, most clay minerals (montmorillonite) are rich with cationic components that determine their cationic exchange capacities. These ions (more especially Na^+ and Ca^{2+}) have high hydrational energies that cause them to be surrounded with many structured units of water during interaction with the water component in a drilling fluid. Formations with high hydrating cations are more likely to swell extensively during interactions with water-based fluids. A drilling fluid with an amino acid component have rich NH^{4+} ions can easily displace and replace the high hydrating Na^+ and Ca^{2+} ions present in the clay's interlayer space (cleavage spacing). This causes the clay to be intercalated with the stabilizer's cationic component that has a lower hydrational energy. This action limits the number of structured units of water surrounding the cations present in the clay, hence, significantly influencing the swelling of the clay. By comparing the two main inhibition mechanisms that natural amino acids employ to stabilize reactive formations, the former which involves maintaining a strong interaction with the base fluid thereby limiting the base fluid's sensitivity to clay minerals is its strongest.

Table 5.2 lists some of the amino acids and derivatives evaluated for their clay stabilizing potentials in drilling fluids.

5.4 UTILIZING ANALYTICAL EVALUATION TECHNIQUES FOR PRESCREENING POTENTIAL AMINO ACIDS FOR DRILLING FLUID-CLAY STABILIZATION APPLICATIONS

With the numerous clarities in terms of the process of formation stabilization via clay stabilizers, developing a prescreening tool that allows one to evaluate the potential of a clay stabilizer prior laboratory and field evaluations is critical. Several experimentations such as the linear swelling test, Fourier transform infrared spectrometer, nuclear magnetic resonance, thermogravimetric analysis, shale recovery test, immersion test, XRD, SEM, zeta potential analysis, contact angle test, and many others are often employed to determine the stabilization potential of a potential clay stabilizer. Although these experimental setups are the most reliable in determining the stabilizing potential of clay stabilizers, they, however, can be time constraining and very costly. The utilization of an analytical tool that can be utilized on-the-fly to evaluate the potential of bio clay stabilizers is very necessary. Klamt[54] introduced a novel solvation model, that is, conductor like screening model for real solvents (COSMO-RS) which has proven to effectively predict the thermodynamic properties of liquid mixtures. Several successful applications of this solvation model have been published in open literature. For example, Sumon and Henni[55] employed COSMO-RS to screen and design ionic liquids for capturing carbon dioxide. Cláudio et al.[56] also utilized

TABLE 5.2
Amino and Poly Amino Acids and Derivatives Evaluated for Their Clay Stabilizing Potentials in Open Literature

Name of Amino Acid / References	Experimental Test Done	Comparing Clay Stabilizers	Other Fluid Composition	Synthesis	pH Studied	Temp (°C)	Field App	Findings
Low-molecular-weight synthetic amino acid (HS) / [14]	Rheology, fluid loss test, cement tolerance, toxicity, swelling test, dispersion test, triaxial test	Polyacrylamide (PHPA)	Fresh water, starch, PAC, lignite, ligno- sulfonate, KOH, soda ash, bentonite, sea salt, lignosulfonate, lignite, barite	NA	7–9	20–65.6	Yes (Wind River Basin of Central Wyoming)	Displayed positive results over PHPA
Oligo poly-l-lysine/ [35,51]	H NMR, GPC, FT-IR, FT-ICR MS, TEM, TGA, XRD, zeta potential, linear swelling test, shale dispersion test	KCl, Jeffamine D230, and EPTAC	$CaCO_3$, white asphalt, polyanionic cellulose, modified starch, and an amphoteric polymer	Thermal copolymerization	7–11	20–180	No	Displayed better results over the conventional clay stabilizers
Poly-l-arginine / [45,51]	H NMR, TGA, FT-IR, linear swelling test, bentonite inhibition test, shale recovery test, microstructural analysis	Polyether diamine, and KCl	$CaCO_3$, white asphalt, polyanionic cellulose, modified starch, and an amphoteric polymer	Thermal copolymerization	7–11	20–180	No	Displayed better results over the conventional clay stabilizers

(Continued)

TABLE 5.2
Continued

Name of Amino Acid / References	Experimental Test Done	Comparing Clay Stabilizers	Other Fluid Composition	Synthesis	pH Studied	Temp (°C)	Field App	Findings
Alanine / [43, 42]	Linear swelling test, rheological test, FTIR, SEM, zeta potential, XRD	KCl, arginine, and proline	Guar gum, ethylene glycol, potassium carbonate	Used without further processing	Basic	20–70	No	Displayed comparable results to KCl
DL Alanine ionic liquids / [52]	TGA, infrared spectroscopy, zeta potential, contact angle, XRD analysis, clay hydration dispersion tests, and anti-swelling rate measurements	KCl	Na bentonite, calcium bentonite, DL-alanine, 98% concentrated sulfuric acid, and ethyl acetate,	DL-alanine undergoes neutralization reaction with concentrated sulfuric acid to generate the target IL in one step	NA	NA	Yes (Qingcheng Oilfield of the Ordos Basin)	All DL Alanine clay stabilizers displayed better results to KCl
Chitosan-grafted l-Arginine / [20]	Linear swelling, shale hot-rolling recovery, shale inhibition durability, and sedimentation experiments, particle size distribution, XRD, SEM, zeta potential analysis, and contact angle	KCl and polyamines (HPA and SIAT)	Na-montmorillonite, Shale (obtained from the Sichuan oilfield in China), xanthan gum, polyanionic cellulose	Synthesized by coupling l-Arg moieties to chitosan	NA	20–150	No	The PCA clay stabilizer displayed better results to the KCl and polyamines clay stabilizers

Poly(citric acid–aspartic acid) copolymer (PCA) / [53]	Fourier transform infrared spectrometer, nuclear magnetic resonance, thermogravimetric analysis, shale recovery test, immersion test, and anti-swelling test	KCl	Sodium bentonite, deionized water, shale cuttings	Citric acid and aspartic acid were mixed into a round-bottom flask and dissolved in 10 mL of dimethylformamide. Conc H_2SO_4 as catalyst. The reaction solution was heated to 180°C. After cooling pH was neutralized using 1 mol NaOH. The purification was operated under reduced pressure distillation.	NA	20–150	No	The PCA clay stabilizer displayed better results to KCl

the solvation model to estimating the hydrogen-bond basicity of ionic liquids. Bavoh et al.[57] employed this model to successfully prescreen potential gas hydrate inhibitors Recently, Al-Maari et al.[58] employed the COSMO-RS thermodynamic tools to predict the efficiency of polar and nonpolar solvents in extracting vegetable oil from rubber seed. Also, Quainoo et al.[43] used the COSMO-RS tool to identify potential amino acids for clay stabilization. They based their approach on the capability of the COSMO-RS model to calculate the excess Gibbs energy, activity coefficient, and hydrogen bond energy of liquid mixtures by focusing on interactions between surface charges of solutes and organic solvents.[42,55] Potent clay stabilizers that operate largely on the mechanism of water invasion reduction can be effectively identified via the output parameters displayed by the model.

5.4.1 Amino Acid Output Parameters and Profiles by COSMO-RS

In relation to clay stabilization, several COSMO-RS thermodynamic outputs are fundamental for determining potential clay stabilizers. They are: (1) sigma (σ) surfaces, (2) activity coefficient, (3) hydrogen bond energy (EHB), and (4) alkyl chain length (surface area). The COSMO-RS sigma surface is just a visual representation that displays the three-dimensional screening charge distribution of molecules by using color spectrums of red, green, and blue. In the σ-surfaces, red represents the hydrogen bonding acceptor surfaces, indicating the regions of the molecules that can accept hydrogen bonds. Green denotes the nonpolar surfaces and regions of the molecules, and blue indicates the hydrogen bonding donor surfaces. Figure 5.3 shows the σ surfaces of six amino acids.

Utilizing these visualizations, several thermodynamic outputs (activity coefficient, hydrogen bond energy, and alkyl chain length) are then computed which can then be utilized to screen the most potent clay stabilizer. The activity coefficient is a thermodynamic parameter in COSMO-RS used to measure the tendency of molecules in a solution to escape. A lower activity coefficient (<1) between the molecules of amino acid and water is an indication of a strong attractive force between the molecules of water and amino acids and vice versa for higher activity coefficient (>1). The hydrogen

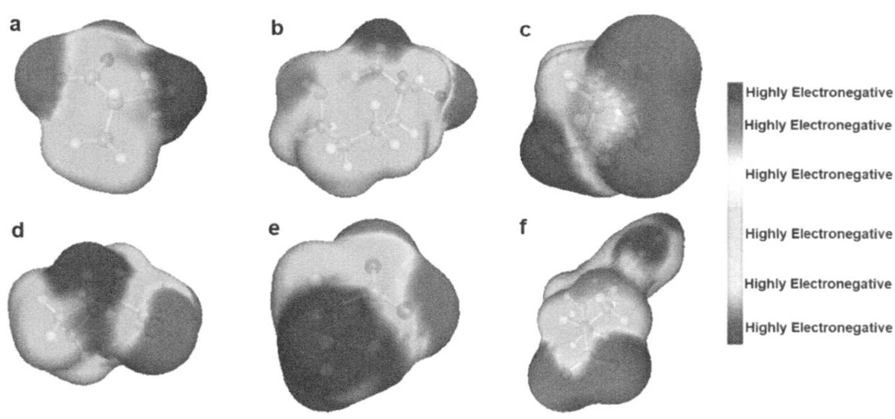

FIGURE 5.3 COSMO-RS sigma surfaces and color codes of several amino acids: (a) alanine, (b) lysine, (c) methionine, (d) proline, (e) glycine, and (h) arginine.

bond energy is defined as the electrostatic force of attraction between molecules with strong dipoles such as amino acids and water. Finally, the alky chain length also known as the surface area of the molecule is the exposed area of the molecule that impacts the molecule's ability to mix with the water molecules. Molecules whose chain lengths are higher are generally considered nonpolar. In terms of stabilizing swelling clays with amino acids, an amino acid that displays strong attractive forces with water (lower activity coefficient), higher electrostatic forces with water, and lower surface areas is likely to display the best clay stabilizing potentials. However, it should be noted that the thermodynamic output displayed by COSMO-RS highly should be confirmed with bench-top experimentations like linear swelling tests. For example, Quainoo et al.[42] evaluated the stabilizing potential of several amino acids via the three COSMO-RS thermodynamic calculated outputs (see Figure 5.4) and performed linear swelling experimentations with the various amino acids.

Figure 5.5 shows the observed inhibition efficiency displayed by the tested amino acids in the presence of a swelling clay and water. Based on the linear swell experiments, the inhibition efficiency follows this order: Glycine > Proline > Threonine > Alanine > Arginine > Methionine > Lysine. Glycine demonstrated the highest potential for inhibiting swelling, while Proline, Threonine, and Alanine showed moderate potential. Arginine, Methionine, and Lysine were the weakest inhibitors.

Comparing both figures, COSMO-RS can only be utilized as an effective tool to screen potential amino acids when all the efficiencies of the outputs (activity coefficient, hydrogen bond energy, and alkyl chain length) were collectively employed. The efficiencies of the COSMO-RS outputs for the various molecules' interactions with

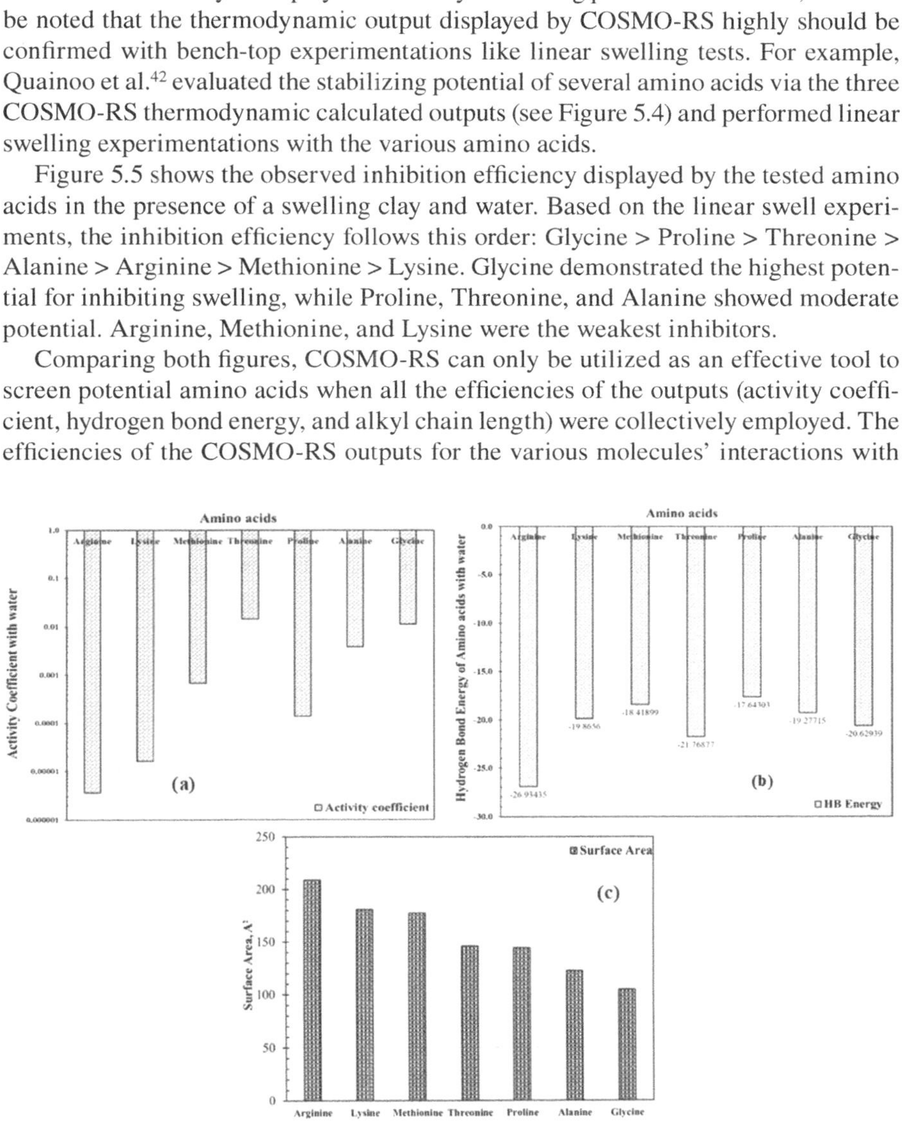

FIGURE 5.4 COSMO-RS outputs parameters of amino acids.

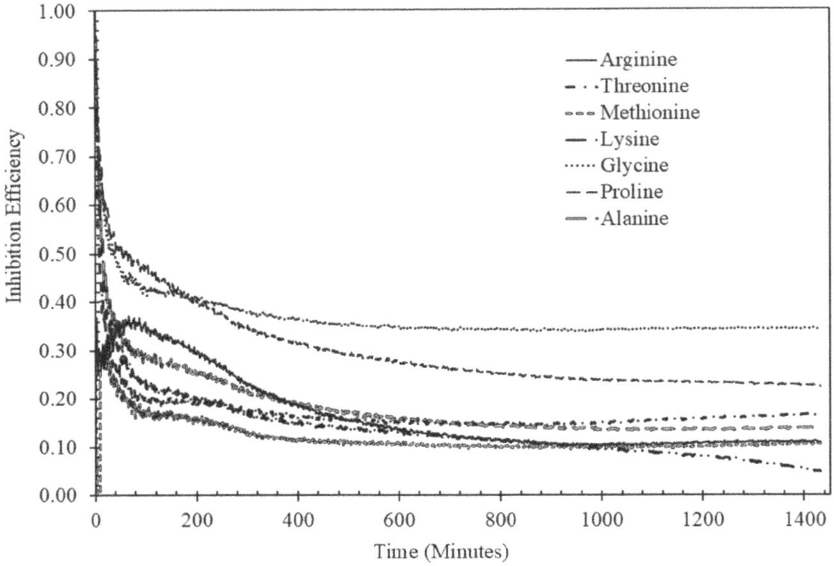

FIGURE 5.5 Swelling inhibition efficiency of amino acids via a Linear Swell Meter.

TABLE 5.3
Efficiencies of COSMO-RS Output Parameters

System Studied	Activity Coefficient Efficiency	Hydrogen Bond Energy Efficiency	Alkyl Chain Length Efficiency	Total COSMO-RS Output Efficiency
Glycine + Water	2.261	3.53	41.05	45.84
Alanine + Water	0.769	3.30	35.14	39.21
Threonine + Water	2.91	3.73	29.47	35.11
Proline + Water	0.028	3.02	29.83	32.88
Methionine + Water	0.135	3.15	24.21	27.50
Lysine + Water	0.003	3.40	23.81	27.22
Arginine + Water	0.001	4.61	20.62	25.23

water were then determined (see Table 5.3). The system with the highest COSMO-RS output efficiency is likely to be the most potent clay stabilizer. See Quainoo et al.[42] for the steps to determine the efficiencies for the various thermodynamic outputs.

Table 5.3 confirms that the COSMO-RS tool was highly effective in determining the most potent clay stabilizer, which is Glycine, as indicated by the linear swelling experiments. The tool also confirmed that Alanine, Threonine, and Proline can be moderately effective clay stabilizers for swelling clays, whereas Methionine, Lysine, and Arginine exhibited very weak potential for stabilizing sensitive clay minerals. Thus, the COSMO-RS tool can be utilized effectively to screen potential chemicals for clay stabilization processes.

5.5 AMINO ACIDS IMPACT IN DRILLING FLUIDS' RHEOLOGY

The rheology of drilling fluids is critical in drilling fluid operations and other processes such as hydraulic fracturing. One of the challenges with the use of KCl for drilling fluids is the negative effect it imposes on drilling fluids. In this section, we report amino acids' impact on the rheological property of drilling fluids. Quainoo et al.[42] examined the rheology of a drilling fluid with amino acids water-based drilling fluid. They formulated a simple drilling fluid comprising guar gum (gelling agent—0.4 wt%), ethylene glycol (dispersant—0.1 wt%), potassium carbonate (pH adjusting agent—0.2 wt%), and water (base fluid). The addition of 2 wt% of either amino acids or KCl to the simple drilling fluid was followed by evaluating the fluid's rheological properties, including (a) flow parameters like yield strength, flow index, and consistency index, and (b) viscoelastic properties assessed through various amplitude and frequency tests. In addition, viscosity-temperature studies were conducted over a range of 25–85°C. The flow parameters of the formulated fluid in the presence of four amino acids are presented in Table 5.4.

A suitable clay stabilizer for the rheology of a drilling fluid is one that maintains the base fluid's flow behavior index ($n < 1$), and the consistency index maintains or slightly improve the yield stress. The inclusion of the amino acids into the drilling fluid did not negatively impact the rheology of the drilling fluids significantly.

Also, the viscoelastic properties of the drilling fluids in the presence of amino acids and KCl clay stabilizers were evaluated (see Table 5.5). This test was done to understand the structural stability of the formulated fluid upon the inclusion of the studied clay stabilizers. Parameters evaluated include the fluid's storage modulus, its loss modulus, its damping factor (tan delta), phase angle, and the strain ranges within which the moduli remained constant (LVER).

Figure 6 shows the system without clay stabilizers displayed higher loss modulus values as compared to its storage modulus. This indicates the formulated fluid is a viscous-dominant one. Upon the inclusion of the amino acids, all except alanine displayed a similar profile with little variations (see Table 5.5 and Figure 5.6). This confirms amino acids' compatibility with the formulated fluid system. On the contrary, alanine showed a significant reduction in the fluid's loss and storage moduli of the formulated fluid. With the storage modulus, Arginine > Glycine > KCl > Proline > Alanine. With the loss modulus, Arginine > Glycine > Proline > KCl > Alanine.

TABLE 5.4
Flow Parameters of Amino Acids in the Presence of a Simple Formulated Drilling Fluid

Studied System	Flow Behavior Index	Consistency Index	Yield Stress
System without clay stabilizers	0.405	1.131	1.014
System + Glycine	0.462	0.734	1.157
System + Alanine	0.425	1.095	2.752
System + Proline	0.440	0.951	2.804
System + Arginine	0.392	1.503	2.255

TABLE 5.5
Viscoelastic Properties of Amino Acids in the Presence of a Simple Formulated Drilling Fluid

Studied System	Storage Modulus, G′	Loss Modulus, G″	Tan delta	LVER
System without clay stabilizers	1.30	1.70	1.30	7.88/1.30
System + KCl	0.84	1.42	1.28	5.93/0.84
System + Glycine	1.01	1.68	1.67	4.44/1.01
System + Alanine	0.34	0.78	2.26	5.93/0.34
System + Proline	0.79	1.5	1.91	10.53/0.79
System + Arginine	1.41	1.80	1.28	5.93/1.41

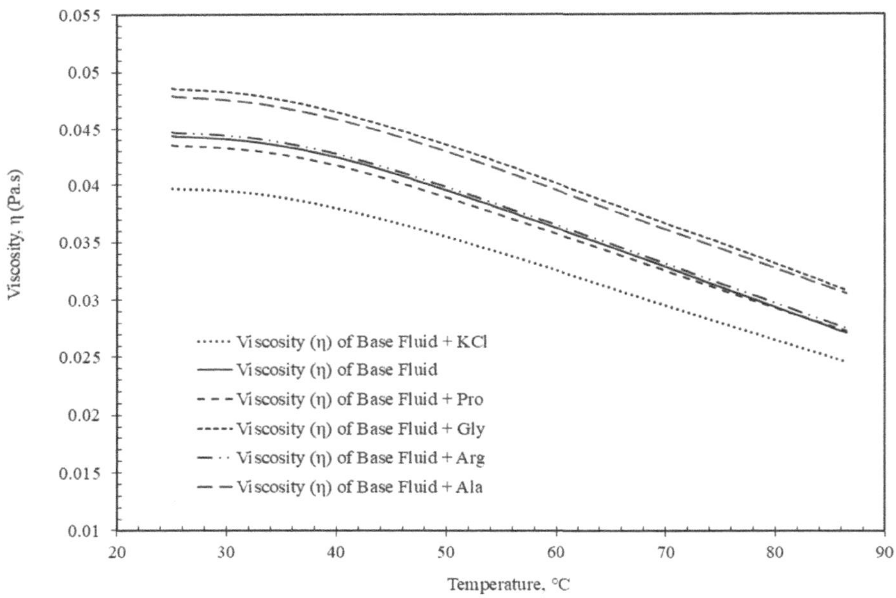

FIGURE 5.6 Temperature profile of amino acid drilling fluids from 25 to 85°C.

A close look at the swelling inhibition potential, the flow parameters, and the viscoelastic properties of the formulated drilling fluids, Glycine still stood out as the most go-to option clay stabilizer.

The tan δ value also known as the fluid's damping factor is obtained by the ratio of G″ to G′. Values >1 indicates the fluid is viscous dominant and elastic-dominant with values <1. With the exception of glycine, all the studied clay stabilizers significantly increased the viscous nature of the formulated fluid. A good clay stabilizer is one that reduces or maintains the damping factor of the drilling fluid. Glycine still stood out.

Finally, the authors also measured the viscosity of the drilling fluid under varying temperatures. Each system was evaluated for 13 min at a shear rate of 200 s^{-1} from 25 to 85°C. Figure 5.6 shows the temperate profiles.

A potent clay stabilizer should be one that maintains or improves the stability of the formulated fluid (regarded as the base fluid). With the exception of KCl and proline, all the studied amino acids improved the viscosity profile of the drilling fluid under increasing temperature in the order: Glycine > Alanine > Arginine > Proline > KCl.

The nondegradation of the viscosity profile as the temperature increases suggests that amino acids such as glycine can also be employed for drilling sensitive clay formations under extreme harsh conditions. This suggests that glycine can be an effective alternative to KCl-stabilized systems for use under any drilling system.

5.6 CONCLUSION AND RECOMMENDATION

5.6.1 Conclusion

This study has demonstrated that amino acids can serve as effective and eco-friendly alternatives to traditional salt-based stabilizers in drilling fluids. The evaluation of the rheological properties, temperature stability, and swelling inhibition of amino acid-stabilized drilling fluids revealed several key findings:

- Amino acids effectively inhibited clay swelling, with glycine showing the highest potential for reducing clay swelling, followed by proline, threonine, and alanine. The ability of amino acids to form strong hydrogen bonds with water molecules reduces the potential for clay minerals to swell, thus improving wellbore stability.
- The inclusion of amino acids such as glycine, alanine, and arginine maintained or improved the rheological properties of the base fluid. These amino acids did not significantly alter the flow behavior index, consistency index, or yield stress, ensuring the fluids' suitability for drilling operations.
- Glycine, alanine, and arginine maintained the viscosity of the drilling fluid across a temperature range of 25–90°C. This indicates that these amino acids can ensure the stability of drilling fluids under varying thermal conditions, unlike proline and KCl, which showed negative impacts on viscosity.
- Oscillatory amplitude tests revealed that the amino acid-stabilized fluids exhibited structural stability within the linear viscoelastic region. This stability is crucial for effective drilling fluid performance in maintaining wellbore integrity.

5.6.2 Recommendation

Based on the findings, the following recommendations are proposed:

a. Given their ability to maintain and enhance rheological properties, structural stability, and swelling inhibition, amino acids such as glycine should be considered for use in drilling fluids, particularly in environmentally sensitive areas.

Extensive field trials should be conducted to validate the laboratory findings and assess the performance of amino acid-stabilized drilling fluids in real-world drilling operations. This will help in understanding their long-term stability and effectiveness under various drilling conditions.

Amino acids offer a green alternative to traditional salt-based stabilizers, aligning with regulatory requirements for environmental protection. The adoption of amino acids can reduce the ecological footprint of drilling operations, making them a preferable choice for sustainable drilling practices.

REFERENCES

1. O'Brien DE, Chenevert ME, JPT. Stabilizing Sensitive Shales With Inhibited, Potassium-Based Drilling Fluids, *J. Pet Technol.* 1973;25:1089.
2. Zhou J, Nasr-El-Din HA. *Proceedings - SPE International Symposium on Oilfield Chemistry* ***2017-April***, 585 (2017).
3. Anderson RL, Ratcliffe I, Greenwell HC, Williams PA, Cliffe S, Coveney PV. *Earth Sci Rev.* 2010;98:201.
4. Himes RE, E.F. Vinson, in *SPE Eastern Regional Meeting* (Society of Petroleum Engineers, 1991).
5. Hu Z, Gao M, Luo Z, Ye Y, Liu Y, Lu L. *Colloids Surf A Physicochem Eng Asp.* 2014;441:685.
6. Koteeswaran S, Habibpour M, Puckette J, Pashin JC, Clark PE. *J Pet Explor Prod Technol.* 2018;8:1273.
7. Qu Y, Lai X, Zou L, Su Y. *Appl Clay Sci.* 2009;3:265.
8. Ren Y, Wang H, Ren Z, Zhang Y, Geng Y, Wu L. Pu X. *Appl Clay Sci.* 2019;182:105248.
9. Soeder DJ. *J Pet Sci Eng.* 2018;163:399.
10. Amorim CLG, Lopes RT, Barroso RC, Queiroz JC, Alves DB, Perez CA, Schelin HR. *Nucl Instrum Methods Phys Res A.* 2007;580:768.
11. Anderson RL, Ratcliffe I, Greenwell HC, Williams PA, Cliffe S, Coveney PV. *Earth Sci Rev.* 2010;98:201.
12. Caenn R, Chillingar GV, Pet J. *Sci Eng.* 1996;14:221.
13. Civan F. *Reservoir Formation Damage.* Gulf Professional Publishing Burlington; 2007.
14. Bruton JR, HC. McLaurine, in *Proceedings - SPE Annual Technical Conference and Exhibition* (Society of Petroleum Engineers, 1993), pp. 127–135.
15. Ismail AR, Jaafar MZ, Sulaiman WRW, Ridzuan A, 8th International Conference on Geological and Geo-Resources Engineering **35**, 24 (2015).
16. Rana A, Arfaj MK, Saleh TA. *Fuel.* 2019;247:237.
17. Quainoo AK, Negash BM, Bavoh CB, Omar T, Tackie-otoo BN. *J Nat Gas Sci Eng.* 2020;79:103380.
18. Ismail I, Ann PH. *J Teknol.* 2009;50:53–65.
19. An Y, Jiang G, Ren Y, Zhang L, Qi Y, Ge Q. *J Pet Sci Eng.* 2015;135:253.
20. Ma J, Yu P, Xia B, An Y, Wang Z. *ACS Appl Bio Mater.* 2019;2:4303.
21. dos L, Cescon S, Quartarone P, da SP, Ribeiro S, Nascimento RSV. *J Appl Polym Sci.* 2018;135:46621.
22. Moslemizadeh A, Samadzadeh Hafshejani K, Shahbazi K, Zaravi Dezfuli M, Zendehboudi S, *Canadian Journal of Chemical Engineering.* 2019;97:384.
23. Barati P, Keshtkar S, Aghajafari A, Shahbazi K, Momeni A. *Petroleum Exploration and Development.* 2016;43:522.
24. Ghasemi M, Moslemizadeh A, Shahbazi K, Mohammadzadeh O, Zendehboudi S, Jafari S. *J Pet Sci Eng.* 2019;178:878.

25. Moslemizadeh A, Khezerloo-ye Aghdam S, Shahbazi K, Zendehboudi S. *J Mol Liq.* 2017;247:269.
26. Aggrey WN, Asiedu NY, Adenutsi CD, Anumah P. *Heliyon.* 2019;5:e01697.
27. Khezerlooe-ye Aghdam S, Moslemizadeh A, Madani M, Ghasemi M, Shahbazi K, Keshavarz Moraveji M. *Chemical Engineering Research and Design.* 2019;147:570.
28. Zhang L, Wu X, Sun Y, Cai J, Lyu S. *Energy Exploration & Exploitation.* 2019;38(2):569–588.
29. Barati P, Shahbazi K, Kamari M, Aghajafari A. *Petroleum.* 2017;3:476.
30. Zhang F, Sun J, Chang X, Xu Z, Zhang X, Huang X, Liu J, Lv K. *Energy & Fuels.* 2019;33:7118.
31. Moslemizadeh A, Shadizadeh SR, Moomenie M. *Appl Clay Sci.* 2015;78:105–106.
32. Shadizadeh SR, Moslemizadeh A, Dezaki AS. *Appl Clay Sci.* 2015;118:74.
33. Li X, Jiang G, Yang L, Peng S. *Colloids Surf A Physicochem Eng Asp.* 2018;539:192.
34. An Y, Yu P. *J Pet Sci Eng.* 2018;161:1.
35. Xuan Y, Jiang G, Li Y, Yang L, Zhang X. *RSC Adv.* 2015;5:84947.
36. Zhong H, Sun D, Huang W, Liu Y, Qiu Z. *Open Fuels and Energy Science Journal.* 2015;8:19.
37. Xuan Y, Jiang G, Li Y, Wang J, Geng H. *Colloids Surf A Physicochem Eng Asp.* 2013;422:50.
38. Zhong H, Qiu Z, Huang W, Cao J. *Appl Clay Sci.* 2012; 36:67–68.
39. Li X, Jiang G, Yang L, Wang K, Shi H, Li G, Wu X. *Energy and Fuels.* 2019;33:9342.
40. Zhong H, Qiu Z, Huang W, Xie B, Wang W. *Open Petroleum Engineering Journal.* 2013;6:49.
41. Khodja M, Khodja-Saber M, Canselier JP, Cohaut N, Bergaya F. *Drilling Fluid Technology: Performances and Environmental Considerations.* Sciyo Publications, London; 2010.
42. Quainoo AK, Borecho C, Opoku K, Alhassan D. *Upstream Oil and Gas Technology.* 2022;9:100080.
43. Quainoo AK, Negash BM, Bavoh CB, Idris A. *J Pet Sci Eng.* 2021;196: 107664.
44. Quainoo AK, Negash BM, Bavoh CB, Idris A, Shahpin HBA, Yaw AD. *Energy & Fuels acs.* 2020; 34(11):13985–14000.
45. Li X, Jiang G, Shen X, Li G. *ACS Sustain Chem Eng.* 2020; 8(4):1899–1907.
46. Quainoo AK, Bavoh CB, Negash BM. *Energy & Fuels.* 2022; 36(7):3539–3548.
47. Low PF, Margheim JF, Soil Sci. Soc. *Am. J.* 1979;473: 473–481.
48. Segad M, Jönsson B, Åkesson T, Cabane B. Langmuir. 2010;26:5782.
49. Peng B, Luo PY, Guo WY, Yuan Q. *J Appl Polym Sci.* 2013;129:1074.
50. Aften CW, Ma Q. Proceedings - SPE International Symposium on Oilfield Chemistry **1**, 639 (2015).
51. Li X, Jiang G, Wang J, Ni X. *J Nat Gas Sci Eng.* 2020;83:103589.
52. Feng L, Bo W, Weichao D, Quande W. 1 *Can. J. Chem. Eng.* 2024; 102(11):3982.
53. Saleh TA, Nur MM, Alarawi AA. *Journal of Physics and Chemistry of Solids.* 2023;181:111455.
54. Klamt A. *Journal of Physical Chemistry.* 1995;99:2224.
55. Sumon KZ, Henni A. *Fluid Phase Equilib.* 2011;310:39.
56. Cláudio AFM, Swift L, Hallett JP, Welton T, Coutinho JAP, Freire MG. *Physical Chemistry Chemical Physics.* 2014;16:6593.
57. Bavoh CB, Lal B, Nashed O, Khan MS, Keong LK, Bustam MA. *Chin J Chem Eng.* 2016;24:1619.
58. Al-Maari MA, Hizaddin HF, Salleh MZM, Hayyan A. *J Mol Model.* 2024;30:1.

6 Amino Acids' Role in Enhanced Oil Recovery

Bennet Nii Tackie-Otoo and
Mohammed Abdalla Ayoub Mohammed

6.1 INTRODUCTION

Enhanced oil recovery (EOR) techniques aim to increase the amount of recoverable oil to approximately two-thirds of the original oil in place, as primary and secondary recovery methods typically extract only about one-third. Primary recovery utilizes the natural energy within the reservoir, often supplemented by artificial lift systems. In contrast, secondary recovery involves the injection of water or gas to maintain reservoir pressure and facilitate the movement of oil toward production wells. Despite these efforts, a substantial portion of oil remains trapped due to capillary forces and unfavorable mobility ratios. This residual oil consists of oil adhering to pore walls after water flooding, as well as bypassed oil.[1]

An effective and widely applied EOR technique for mobilizing trapped oil and improving displacement efficiency is chemical flooding.[2] This method involves injecting chemical agents such as alkalis, surfactants, polymers, and nanoparticles.[3] Among these, surfactants have demonstrated the highest effectiveness, primarily by lowering interfacial tension (IFT) and modifying reservoir wettability to enhance oil recovery.

Surfactant flooding and other combined techniques involving surfactants have been widely validated as both feasible and effective through laboratory experiments and field applications.[4] Surfactants, also known as surface-active agents, are compounds that self-assemble into molecular aggregates called micelles in solution and adsorb at the interface between a solution and another phase. These characteristics stem from their amphiphilic nature, consisting of a hydrophilic (water-soluble) or polar head group and a hydrophobic (oil-soluble) or nonpolar tail. The hydrophilic head typically contains functional groups such as carboxylates, sulfates, sulfonates, carboxybetaines, sulfobetaines, polyoxyethylenated chains, or quaternary ammonium salts. Meanwhile, the hydrophobic tail is generally composed of an alkyl chain with 8–22 carbon atoms, which may be either branched or linear.[5]

Significant research efforts have been dedicated to developing surfactants optimized for specific applications.[1] However, as observed in the pharmaceutical and cosmetic industries, a major concern associated with industrial surfactants is their biodegradability and biocompatibility, given their potential environmental impact, particularly on aquatic ecosystems.[6,7] In recent years, natural and bio-based surfactants have gained increasing attention for their environmentally friendly properties, including biodegradability, biocompatibility, and excellent surface activity.[8,9] Studies

DOI: 10.1201/9781003502388-6

have demonstrated that plant-derived anionic surfactants synthesized from vegetable oils exhibit outstanding surface-active properties and strong potential for EOR.[10–16]

Certain amino acids, including tyrosine, tryptophan, lysine, and arginine, exhibit amphipathic properties due to the presence of hydrophobic alkyl chains and charged head groups. This dual characteristic enables them to simultaneously interact with both polar and nonpolar surface ligands, positioning them as potential candidates for surfactant applications.[17]

Watry and Richmond[18] employed vibrational sum-frequency spectroscopy to investigate the orientation and conformation of amino acids within monolayers adsorbed at an oil/water interface.[19] Their findings revealed that amino acids adopt distinct conformational orientations depending on the nature of the interface, whether polar or nonpolar. At a nonpolar decane-water interface, amino acids with nonpolar side chains, such as methionine and tryptophan, align themselves with their hydrophobic side chains facing the interface, while their polar amine ($-NH_2$) and carboxyl ($-COOH$) groups remain directed toward the bulk aqueous phase. Conversely, amino acids with ionic side chains, such as L-arginine and L-lysine, tend to orient their side chains horizontally along the interface due to their charged nature.

Despite these insights into the behavior of amino acids at oil/water interfaces, there are limited studies on their application as chemical agents for EOR. Amino acids' ability to adsorb at interfaces and modify surface properties makes them potentially valuable for EOR processes, but further research is needed to explore their effectiveness and optimize their use in this context. This study highlights the importance of molecular structure and interfacial properties in determining the behavior of amino acids, which could have broader implications for their use in industrial applications, including corrosion inhibition and EOR.

Orozco et al.,[20] inspired by the wettability alteration mechanisms of low salinity water flooding, explored amino acids as innovative wettability modifiers for carbonate reservoirs. Their study focused on assessing the impact of glycine, the simplest amino acid, and β-alanine on the wettability modification of oil-aged calcite in a brine solution. Similarly, Atta et al.[21] explored the potential of amino acids, including L-arginine, L-lysine, L-tryptophan, and L-methionine, in reducing IFT and modifying rock surface wettability.

The limited research and application of amino acids as EOR agents may be due to their short alkyl chain lengths, a crucial property of surfactants that significantly influences their surface and interfacial behavior. To address this limitation, amino acids have been effectively utilized as feedstock for synthesizing bio-based surfactants, enhancing their performance in EOR applications.

6.2 BIO-BASED SURFACTANTS

Because of their high biodegradability, environmental compatibility, and excellent surface properties, a new class of surfactants derived from renewable raw materials has garnered significant attention. These include α-sulfofatty acid methyl esters (SMEs), acylated amino acids categorized as amino acid-based surfactants (AASs), and Nopol alkoxylates.[22] In EOR research, SMEs synthesized from various vegetable oils have exhibited remarkable surface activity and a strong potential for enhancing oil recovery.[23–29] In addition, bio-based zwitterionic surfactants derived from vegetable oil fatty acids have been proposed for EOR applications.[30–34]

Acylated amino acids, a subclass of AASs, are fatty acids modified with an amido linkage in their hydrocarbon chains. This structural modification improves their solubility, adsorption behavior, and stability in hard water and high-electrolyte environments compared to unmodified fatty acids.[35] Despite their potential, AASs have received limited research attention in EOR applications.[10–13,15,16] While their performance has yet to match that of conventional surfactants, future studies could focus on optimizing their molecular design to improve their EOR efficiency. Notably, sodium salts of N-α-acylamino acids and fatty acid mixtures have demonstrated comparable or even superior performance to petroleum-based surfactants.[36] Recent findings indicate that these AASs hold promise for EOR applications.[37–39]

TABLE 6.1
Bio-Based Surfactants for EOR Application

Surfactant	Source	Reference
Sodium methyl ester sulfonate (J-SMES)	Jatropha oil	27,40
Sodium methyl ester sulfonate (C-SMES)	Castor oil	41,42
Sodium methyl ester sulfonate (S-SMES)	Soap-nut	25,43
Sodium ethyl ester sulfonate (P-SEES)	Palm oil	29
Sodium ethyl ester sulfonate (C-SEES)	Coconut oil	26
N-phenyloctadecanoicamdopropyl-N, N-dimethylcarboxylbetaine (POAPMB)	Oleic acid	30
N-tolyloctadecanoicamidopropyl-N, N-dimethylcarboxylbetaine (TOAPMB)		
N-ethylphenyloctadenoicamidopropyl-N, N-dimethylcarboxylbetaine (EOAPMB)		
N-phenyloctadecanoicamidoethyl-N, N-dimethylcarboxylbetaine (POAEMB)		
N-phenyloctadecanoicamidoethyl-N, N-diethylcarboxylbetaine (POAEEB)		
N-phenylfattyamidopropyl-N, N-dimethylcarboxylbetaine (CPDB)	Castor oil	32,33
N-phenylfattyamidopropy$_L$-N, N-dimethylcarboxylbetaine (WPDB)	Waste cooking oil	34
Bio-based sulfonic zwitterionic surfactant (BSZS)	Soybean oil	31
N-hexadecanoicamidopropyl-N, N-dimethylcarboxylbetaine (HDB)	Palmitic acid	32
N-octadecanoicamidopropyl-N, N-dimethylcarboxylbetaine (ODB)	Stearic acid	
N-phenyllinoleicamidopropyl-N, N-dimethylcarboxylbetaine (PLDB)	Linoleic acid	
N-phenylricinoleicamidopropyl-N, N-dimethylcarboxylbetaine (PRDB)	Ricinoleic acid	
N-phenyllinolenicamidopropyl-N, N-dimethylcarboxylbetaine (PLDB')	Linolenic acid	
N-phenylfattyamidopropyl-N, N-dimethylcarboxyl betaine (GPDB)	Soybean oil	
N-phenylfattyamidopropyl-N, N-dimethylcarboxyl betaine (CPDB')	Cottonseed oil	
N-phenylfattyamidopropyl-N, N-dimethylcarboxyl betaine (RPDB)	Rapeseed oil	
Diacyl amino acid surfactant (DAAS)	Lauric acid / lysine	12

(Continued)

TABLE 6.1
(Continued)

Surfactant	Source	Reference
(S)-2-amino-6-dodecanamidohexanoic acid (lauroyl lysine/L-Lys)	Lauric acid / lysine	10,16
(S)-2-dodecanamido-5-guanidinopentanoic acid (lauroyl arginine/L-Arg)	Lauric acid / arginine	11,13,15
Lauroyl cysteine (L-Cys)	Lauric acid / cysteine	13
(S)-dodecanoyl-L-aspartic acid (lauroyl aspartic acid/L-Asp)	Lauric acid / aspartic acid	44
Sodium cocoyl alaninate (SCA)	Coconut oil / alanine	37,38
N-lauroyl sarcosine (NLS)	Fatty acid / sarcosine	39
N-lauroyl-L-glutamic acid (LGA)	Fatty acid / glutamic acid	39

Further exploration and optimization of bio-based surfactants for EOR applications could lead to sustainable, high-performance alternatives to petroleum-derived surfactants in Table 6.1.

6.2.1 Ecotoxicology of Bio-Based Surfactants

This chapter explores ecotoxicological studies on bio-based surfactants to assess their potential in mitigating the environmental impact of surfactant flooding processes. While extensive research has been conducted on the EOR capabilities of bio-based surfactants, ecotoxicological evaluations remain scarce in the literature, except for studies on bio-based zwitterionic surfactants. The assumption of their environmental friendliness is primarily based on their derivation from renewable resources. Therefore, this discussion focuses on ecotoxicological assessments of bio-based surfactants, particularly regarding their applications across various industries.

Ishak et al.[45] conducted a comprehensive study on the ecotoxicology of methyl ester sulfonate (MES) homologues derived from palm oil. Their findings indicated that all MES samples were readily biodegradable, with each homologue achieving over 60% biodegradability within 28 days. In addition, they observed that the C_{12} and C_{14} homologues exhibited no toxicity toward *P. subcapitata*. However, as the alkyl chain length increased, toxicity levels rose while biodegradability declined.

The biodegradability of bio-based zwitterionic surfactants has also been investigated in studies evaluating their potential for EOR.[30,32–34] Zhang et al.[30,33,34] and Gang et al.[32] reported that bio-based carboxylbetaine zwitterionic surfactants exhibit ultimate biodegradability over several months, making them viable for EOR applications. Zhang et al.[33] further compared the biodegradability and toxicity of N-phenylfattyamidopropyl-N, N-dimethylcarboxylbetaine (CPDB) with alkylbenzene sulfonate (ABS) and alkyl naphthalene sulfonate (ANS), revealing that CPDB demonstrated biodegradability comparable to ABS but slightly higher than ANS, while exhibiting lower toxicity than both.

Similarly, AASs have been widely studied for their biodegradability, with findings indicating their rapid breakdown into amino acids and fatty acids.[46] Akinari et al.[47] reported microbial degradation rates of 57–73% within 14 days for synthesized AASs. Brito et al.[48] noted that AASs in vesicle-containing mixtures exhibited lower ecotoxicity compared to the conventionally used dodecyltrimethylammonium bromide. Furthermore, Rosa et al.[49] found that cationic AASs appeared to be nontoxic relative to traditional cationic surfactants. For a more comprehensive analysis of AAS biodegradability and toxicity, readers are referred to literature.[46]

The ecotoxicological studies reviewed suggest that various bio-based surfactants could help mitigate environmental concerns associated with surfactant flooding processes. However, existing MES and AAS studies are primarily based on applications in other fields where surfactant concentrations are lower than those used in EOR. Since ecotoxicological properties can vary with concentration, further research is recommended to assess the environmental impact of bio-based surfactants under EOR-relevant conditions.

6.3 AAS AND OTHER BIO-BASED SURFACTANTS AS EOR AGENTS

The choice of surfactant suitable for EOR depends on its ability to reduce IFT between crude oil and brine, along with properties such as thermal stability, tolerance to brine salinity and hardness, solubility in brine, phase behavior characteristics, adsorption under both static and dynamic conditions, and recovery efficiency under reservoir conditions.[4] Surfactants are integral to EOR, primarily by reducing IFT between immiscible phases, such as oil and brine. Lowering IFT facilitates the mobilization of residual oil and promotes oil bank formation. By mitigating capillary forces that would otherwise trap oil, surfactants enhance displacement efficiency.[50] Among the surfactants explored for EOR, AASs and other bio-based alternatives have emerged as promising candidates due to their ability to reduce IFT, alter wettability, and improve emulsification, thereby enhancing oil recovery. This section reviews key findings from studies investigating the potential of bio-based surfactants in chemical EOR (cEOR) applications.

6.3.1 Surface Activity and Aggregation Behavior in Aqueous Solution

Surfactant surface activity increases with concentration until reaching the critical micelle concentration (CMC), at which point monomeric surfactants aggregate into micelles. This transition reduces the availability of free monomers, subsequently limiting further surface activity. Understanding surfactant aggregation behavior, particularly CMC determination, is essential for optimizing their performance in EOR applications.

Research on bio-based surfactants has focused on CMC determination, yet studies on their surface activity and bulk-phase aggregation behavior remain limited. Further investigations are necessary to elucidate how structural properties and microenvironmental factors influence their efficiency in EOR applications. In addition, micelle formation plays a critical role in the creation of microemulsions,[51] which are vital for improving oil recovery with surfactants. In this section, we discuss the surface activity and aggregation behavior of bio-based surfactants, followed by a focused discussion on studies of AASs for EOR applications. There are different methods for determining CMC. However, the surface tension (Figure 6.1a) and the conductivity (Figure 6.1b) and surface tension methods have been widely employed in studies of bio-based surfactants.[24–27,33,43,44] In addition, techniques such as pH (Figure 6.1c) and turbidity (Figure 6.1d) measurements have also been utilized.[10,11,13] The CMC can also be determined using IFT measurements (Figure 6.2). However, when crude oil is involved, its various components can influence the surfactant's CMC. For instance, the CMC of sodium cocoyl alaninate (SCA) is approximately 0.2 wt% (2,000 ppm) but was observed to be around 0.15 wt% (1,500 ppm) when IFT measurements between the surfactant aqueous solution and crude oil were used to determine the CMC.[37] This discrepancy is attributed to crude oil components such as asphaltenes, organic acids, resins, and solids, which can significantly affect the solubility of certain polar organic compounds in both oleic and aqueous phases media.

A wide range of CMCs has been reported for bio-based surfactants, spanning several orders of magnitude, from as low as 1×10^{-3} mM to over 10 mM (~0.7–9,000 ppm). The primary determinant of CMC variation in surfactants is their molecular structure.[22] Notably, the hydrophobic group exerts a greater influence on CMC than the hydrophilic group. Since plant-based oils contain diverse triglyceride compositions, the fatty acids derived from them also vary. Consequently, bio-based

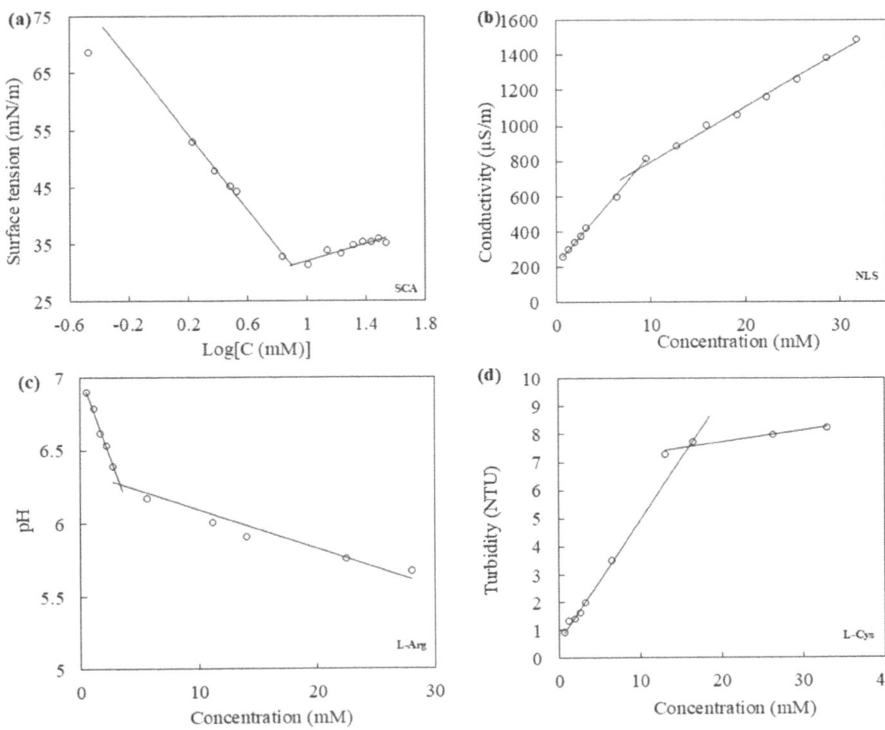

FIGURE 6.1 CMC of AASs determined using (a) surface tension measurements[37], (b) conductivity measurements[37], (c) pH measurements[13] and (d) turbidity measurements.[13]

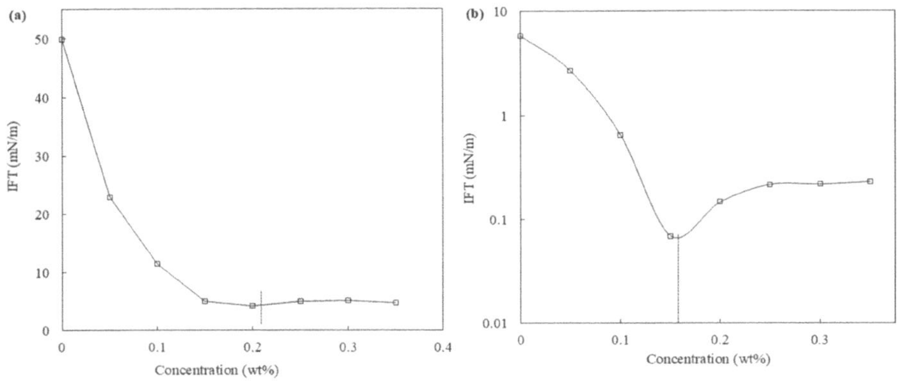

FIGURE 6.2 CMC of SCA determined from (a) aqueous-decane IFT[37] and (b) aqueous-crude oil IFT measurements[38].

surfactants consist of hydrophobic chains with different saturated and unsaturated fatty acids, depending on the fatty acid source.

This structural variability explains why the same type of surfactant, when derived from different plant oils, can exhibit distinct CMC values, as illustrated in Table 6.2. A comparison of reported CMCs for bio-based surfactants (Table 6.2) reveals that their variations align with those observed in conventional surfactants. The following key observations hold for bio-based surfactants[22]:

- Hydrophobic Chain Length: Longer hydrophobic chains result in lower CMC values. Incorporating a phenyl group into the hydrophobic chain reduces the CMC by an amount equivalent to approximately 3.5 methylene groups.
- Unsaturation: The presence of carbon-carbon double bonds in the hydrophobic group typically increases the CMC compared to its saturated counterpart.
- Nonionic and zwitterionic surfactants generally possess longer hydrophobic chains compared to ionic surfactants.
- For surfactants with the same alkyl chain length, nonionic surfactants exhibit lower CMCs than zwitterionic surfactants, while zwitterionic surfactants have lower CMCs than ionic surfactants. Among ionic surfactants, anionic surfactants generally display lower CMCs compared to cationic surfactants.
- The presence of bulky groups either in the hydrophilic or hydrophobic group elevates the CMC.

The determination of the CMC through surface tension measurement is considered highly versatile, as it also provides insights into the adsorbed layer at the air-water interface.[52] This information is crucial for assessing both the efficiency of surface tension reduction by a surfactant and its adsorption effectiveness. The ability of surfactants to lower IFT and alter wettability key factors in their EOR applications primarily depends on their adsorption efficiency.

TABLE 6.2
CMC and Structure of Bio-Based Surfactants

Surfactant	CMC (ppm)	Structure of Surfactant	Reference
J-SMES	8,000	$R{-}CH(SO_3Na)COOCH_3$	27,40
C-SMES	4,500	$H(CH_2)_6CH(OH)CH_2CH{=}CH(CH_2)_6CH(SO_3Na)COOCH_3$	41,42
S-SMES	9,000	$R{-}CH(SO_3Na)COOCH_3$	25,43
P-SEES	8,000	$R{-}CH(SO_3Na)COOCH_2CH_3$	29
C-SEES	5,000	$R{-}CH(SO_3Na)COOCH_2CH_3$	26
POAPMB	2.81	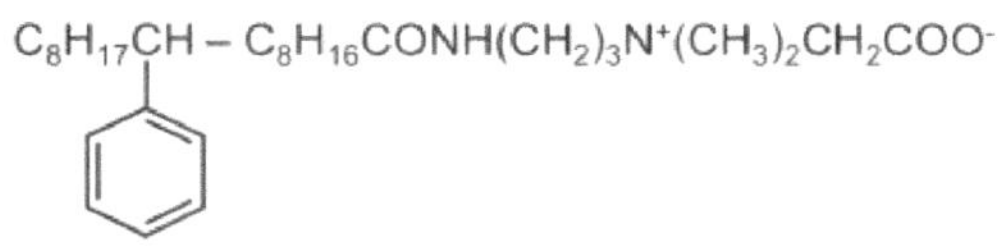	30
TOAPMB	2.23	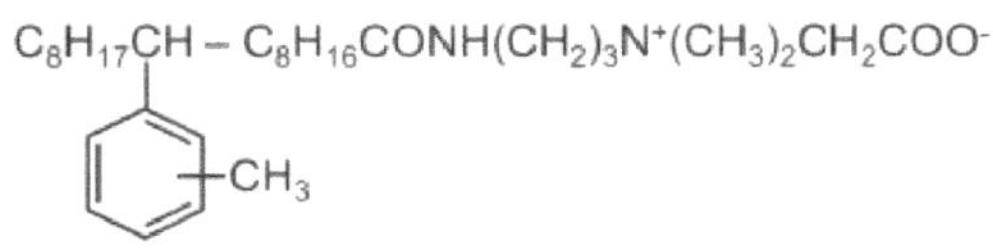	30
EOAPMB	1.76	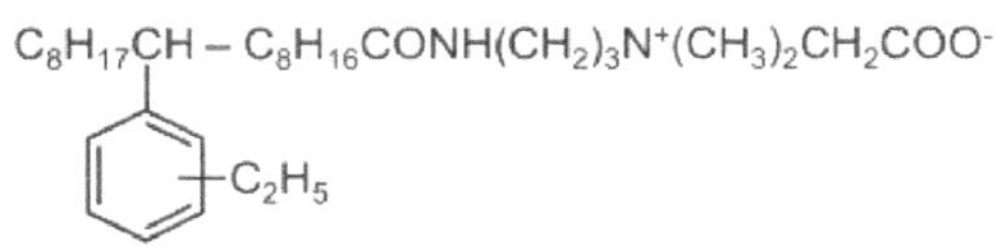	30
POAEMB	1.99	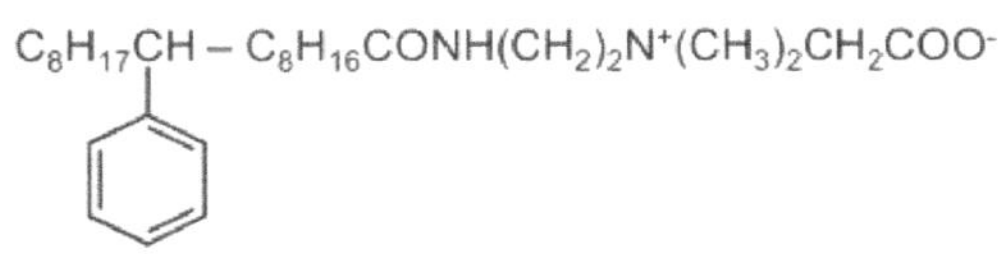	30
POAEEB	1.62	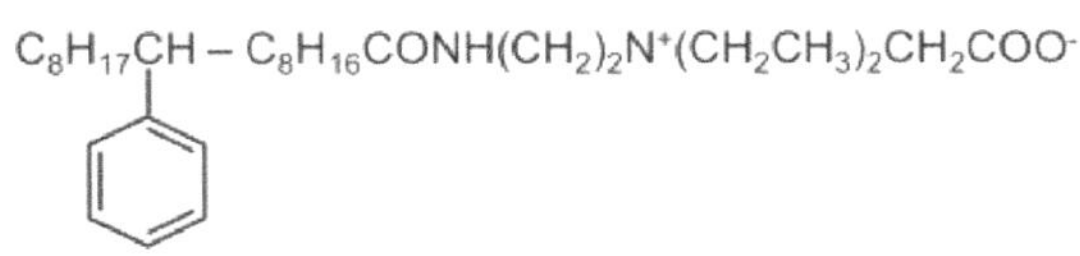	30

(Continued)

TABLE 6.2
(Continued)

Surfactant	CMC (ppm)	Structure of Surfactant	Reference
CPDB	3.52	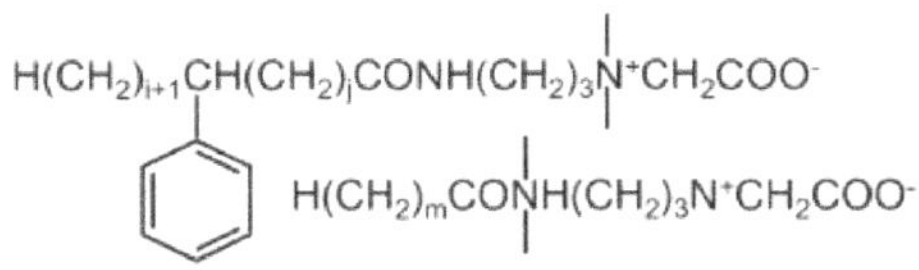	32,33
WPDB	0.734	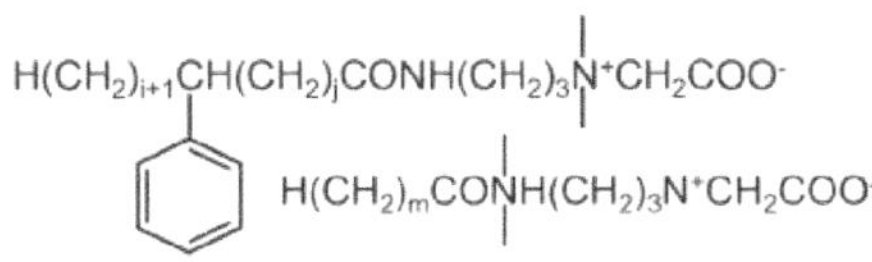	34
BSZS	33.34	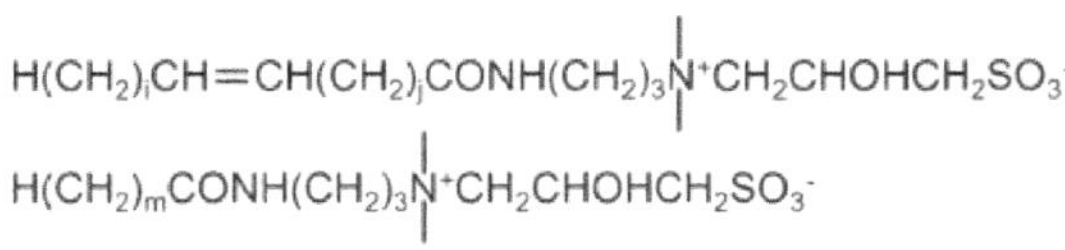	31
HDB	11.1	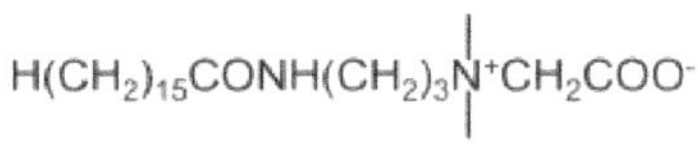	32
ODB	10.1	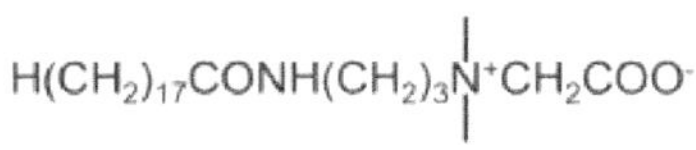	32
PLDB	3.0	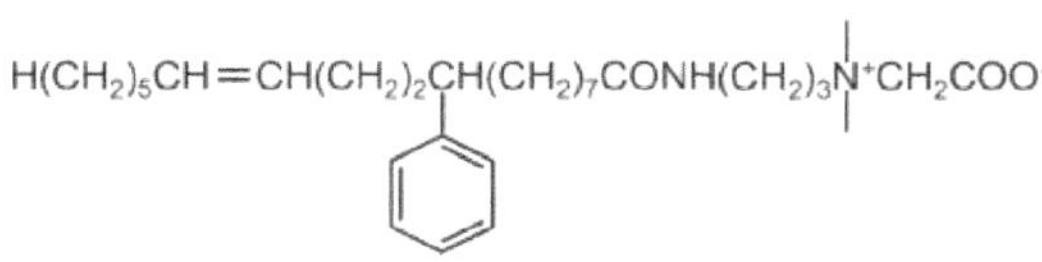	32
PRDB	5.1	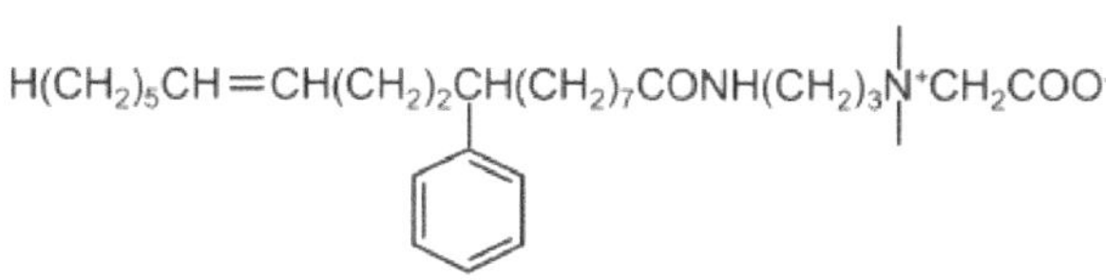	32
PLDB′	2.7	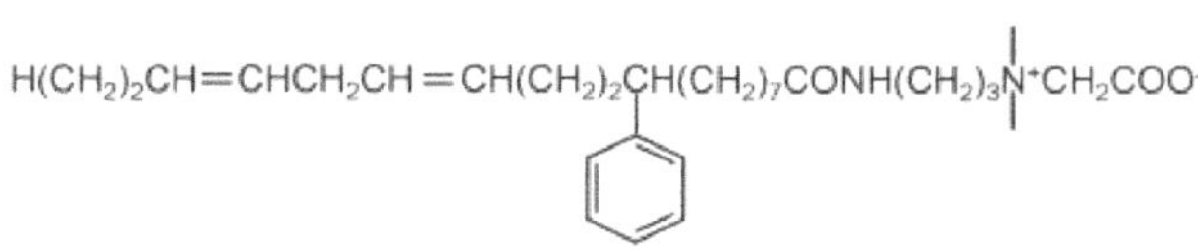	32
GPDB*	7.9	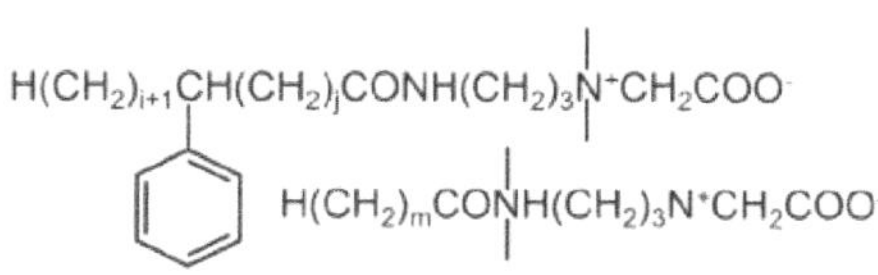	32

(Continued)

TABLE 6.2
(Continued)

Surfactant	CMC (ppm)	Structure of Surfactant	Reference
CPDB*	7.8	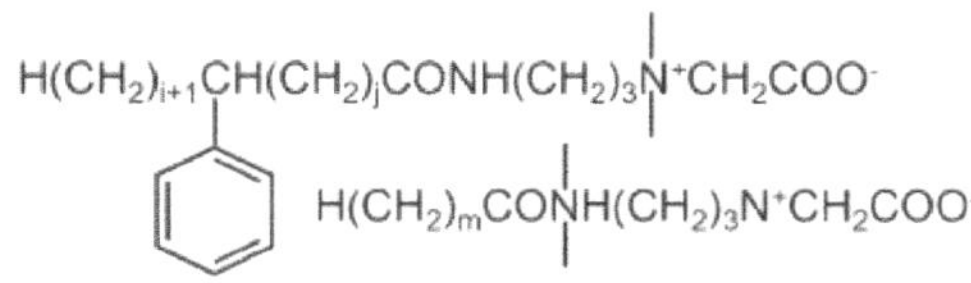	32
CPDB'*	4.2	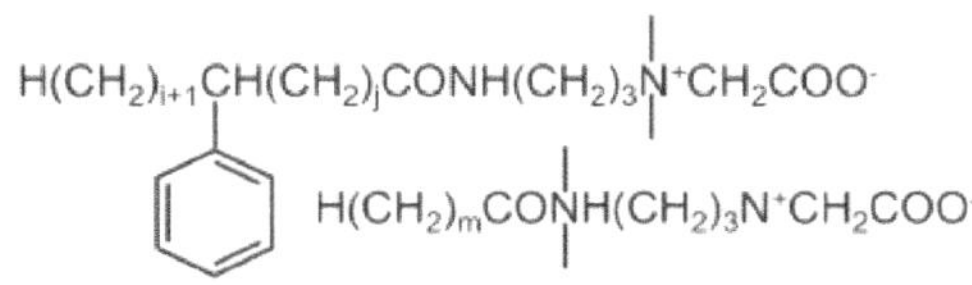	32
RPDB*	6.3	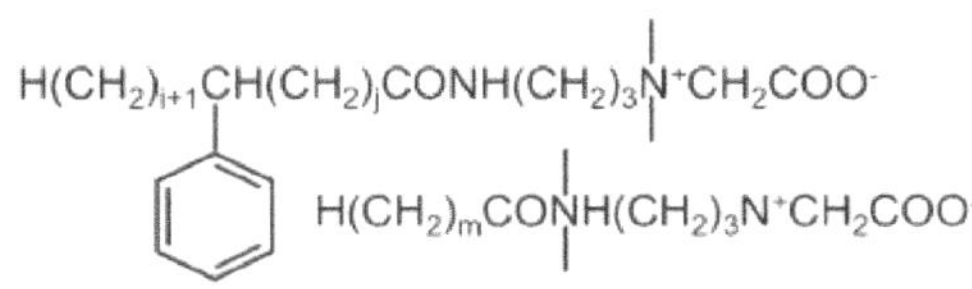	32
L-Lys	5,110	$H(CH_2)_{11}CONH(CH_2)_4CH(NH_2)COOH$	10,16
L-Arg	9,100	$H(CH_2)_{11}CONHC(COOH)(CH_2)_3NHC(=NH)-NH_2$	11,15
L-Arg	2,000	$H(CH_2)_{11}CONHC(COOH)(CH_2)_3NHC(=NH)-NH_2$	13
L-Cys	4,500	$H(CH_2)_{11}CONHC(COOH)CH_2SH$	13
L-Asp	2,000	$H(CH_2)_{11}CONHC(COOH)CH_2COOH$	44
SCA	2,330	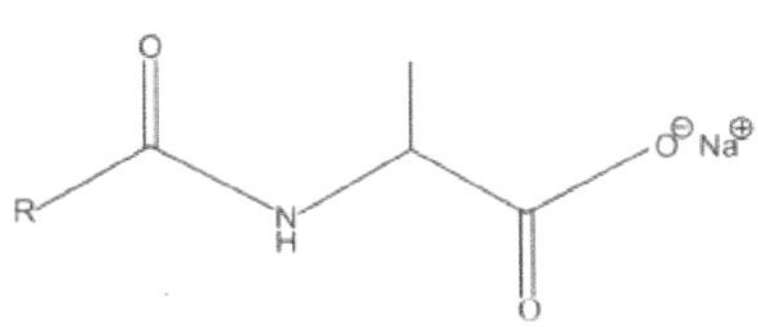	37,38

(Continued)

TABLE 6.2
(Continued)

Surfactant	CMC (ppm)	Structure of Surfactant	Reference
NLS	1,852	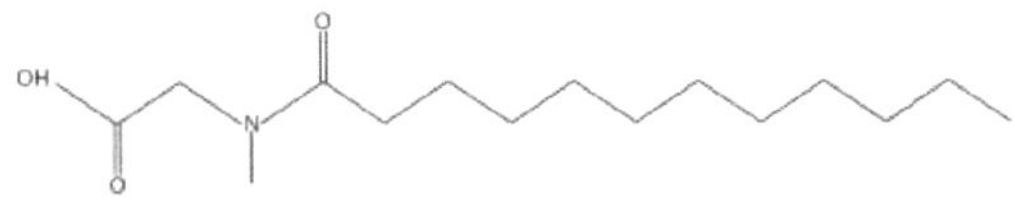	39
LGA	11,998	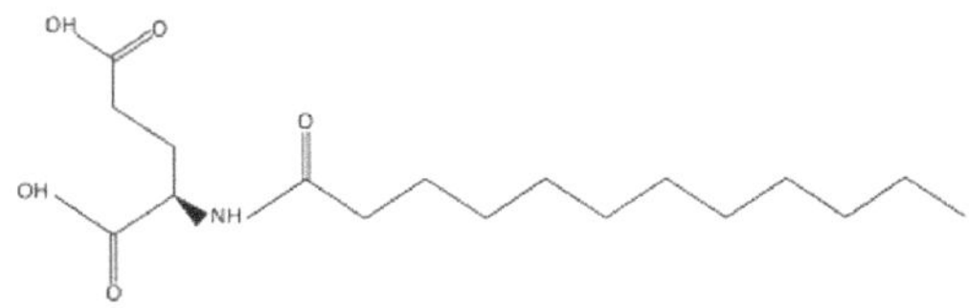	39

* Simulated fatty acid mixture

The parameter pC_{20} represents the bulk-phase concentration required to achieve a specific reduction in surface tension, serving as a measure of a surfactant's efficiency in lowering surface tension. In contrast, effectiveness is indicated by the surface pressure at the CMC (ΠCMC), which reflects the maximum achievable surface tension reduction, independent of bulk-phase concentration.[53] The extent of surfactant adsorption at the interface is quantified by the surface excess concentration (Γ_m) and the minimum surface area per molecule at surface saturation ($a^s{}_m$), both of which are determined using the Gibbs adsorption isotherm.[22] These parameters are mathematically expressed as follows:

$$pC_{20} = -\log C_{20} \tag{6.1}$$

$$\Pi_{CMC} = \gamma_0 - \gamma_{CMC}, \tag{6.2}$$

$$\Gamma_m = -\frac{1}{2.303nRT}\left(\frac{d\gamma}{d\log C}\right), \tag{6.3}$$

$$a_m^s = \frac{1}{N\Gamma_m}, \tag{6.4}$$

where C_{20} represents the bulk phase surfactant concentration required to reduce the surface tension of a pure solvent by 20 mN/m. Here, γ_0 denotes the surface tension of the pure solvent, and γ_{CMC} is the surface tension of the surfactant at the CMC. In Equations 6.3 and 6.4, n represents the number of solute species whose interfacial concentration changes with variations in the bulk phase concentration, R is the universal gas constant (8.314 $J{\cdot}K^{-1}{\cdot}mol^{-1}$), T is the absolute temperature in Kelvin, and N is Avogadro's number.

Thermodynamic properties of micellization and adsorption can also be derived from electrical conductivity data. Key parameters for this analysis include the Gibbs free energy of micellization (ΔG°_{mic}) and the Gibbs free energy of adsorption (ΔG_{ads}). This thermodynamic analysis helps elucidate how structural and environmental factors influence micellization and adsorption processes.[22] Using the CMC, ΔG°_{mic} can be calculated using Equation (6.5)[54]:

$$\Delta G^{\circ}_{mic} = (1+\beta)RT\ln X_{CMC} \tag{6.5}$$

The CMC can also be expressed in terms of mole fraction (X_{CMC}), while the degree of counterion binding (β) of the micelle is determined by its degree of ionization (α). The degree of ionization is obtained from the ratio of the slope before the CMC to the slope after the CMC in a conductivity versus concentration plot, with the relationship defined as $\beta = 1 - \alpha$.[55] In addition, the process of surfactant adsorption at the interface can be evaluated through the standard free energy of adsorption (ΔG°ads), which is derived from the standard free energy of micellization (ΔG°mic) using Equation (6.6)[56]:1

$$\Delta G^{\circ}_{ads} = \Delta G^{\circ}_{mic} - \frac{\Pi_{CMC}}{\Gamma_m} \tag{6.6}$$

The majority of studies on bio-based surfactants for EOR applications have not included this type of investigation. However, analyses were conducted by Pal et al.[24] and Saxena et al.[43] for J-SMES and S-SMES, respectively. In addition, Zhang and Cai[33] examined the adsorption efficiency of CPDB using surface tension data. For AASs as EOR agents, similar studies have been performed for SCA, N-lauroyl sarcosine (NLS), and N-lauroyl-L-glutamic acid (LGA).[37,39]

As summarized in Table 6.3, LGA is the least effective in reducing surface tension among these AASs. This is attributed to increased electrostatic repulsion in ionic surfactants, particularly due to the additional carboxylate group in LGA's head group, which further diminishes its efficiency. In contrast, SCA and NLS exhibit comparable surface tension reduction efficiency, both outperforming sodium dodecyl sulfate (SDS). The superior performance of SCA is linked to its mixed acyl chains, while NLS benefits from its organic counterion.[57]

Among these AASs, SCA is the most effective in reducing surface tension. Although LGA demonstrates strong surfactant adsorption due to the presence of an

TABLE 6.3
Parameters Derived from Surface Tension Data at 25°C

Surfactant	CMC (mM)	Υ_{cmc} (mN/m)	pC_{20}	CMC/C_{20}	Π_{cmc} (mN/m)	Γ_m (μmol/m²)	a^s_m (Å²)
SCA	7.94	31.21	2.74	4.32	40.79	2.86	58.15
NLS	5.9	40.4	2.75	3.3	31.6	1.99	83.4
LGA	32.2	42	1.67	1.5	30	3.26	50.83
SDS	7.94	39.6	2.51	2.6	32.5	3.16	52.55

TABLE 6.4
Thermodynamic Parameters Derived from Conductivity Data

Surfactant	CMC (mM)	α	β	ΔG°_{mic} (kJ/mol)	ΔG°_{abs} (kJ/mol)
SCA	9.5	0.62	0.38	−29.70	−43.98
NLS	8.6	0.51	0.49	−32.46	−48.33
LGA	26.6	0.60	0.40	−26.45	−35.64
SDS	8.7	0.44	0.56	−33.91	−44.20

amide bond in its head group enabling self-assembly through intermolecular hydrogen bonding SCA and NLS achieve greater surface tension reduction due to their enhanced adsorption properties.[58]

A comparison of the thermodynamic parameters for AASs in Table 6.4 reveals that the micellization and adsorption processes of LGA are the least spontaneous. As previously discussed, the introduction of a carboxylate group in LGA induces electrical repulsion during aggregation in both the interface and bulk phase, leading to more positive free energy values.[59] Consequently, both micellization and adsorption are less spontaneous. Conversely, the micellization and adsorption of NLS are more spontaneous than those of SCA due to the presence of an organic counterion.[57] Compared to SDS, SCA exhibits lower spontaneity in both micellization and adsorption at the air/water interface. However, an analysis of the difference between these two parameters for SCA and SDS indicates that the relative spontaneity of surfactant adsorption at the air/water interface compared to micellization is higher for SCA than for SDS. Furthermore, while NLS demonstrates lower spontaneity in micellization relative to SDS, it exhibits greater spontaneity in adsorption.

Microenvironmental factors significantly influence the aggregation behavior of surfactants, thereby affecting their CMCs. The effects of added electrolytes and temperature on the CMC of AASs have been documented in the literature. For instance, the CMC of SCA initially increases with temperature up to 308.15 K and then decreases as the temperature rises to 338 K (Figure 6.3a), following a trend similar to that of sodium n-dodecanoyl sarcosinate.[55] This variation in CMC with temperature arises from two opposing effects: reduced hydration of the hydrophilic headgroup promotes micelle formation, while the disruption of structured water around the hydrophobic tail hinders micelle formation.[22] Thus, the observed trend in CMC with increasing temperature depends on the balance between these two competing phenomena.

An increasing trend in CMC of lauroyl lysine with salinity is reported in Rostami et al.[16] As a zwitterionic surfactant, lauroyl lysine exhibits amphoteric behavior in a neutral medium.[46] The addition of electrolytes, which function as water structure breakers, induces a salting-in effect on the hydrophobic region of surfactant monomers, leading to an increase in CMC. In contrast, Madani et al.[15] observed a decreasing trend in the CMC of lauroyl arginine with increasing salinity. Since lauroyl arginine is an anionic surfactant, the presence of electrolytes compresses the ionic atmosphere surrounding the headgroup, reducing electrostatic repulsion among

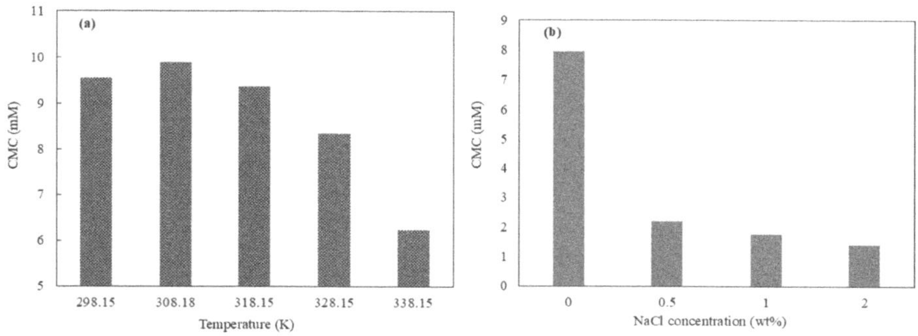

FIGURE 6.3 CMC variation with (a) temperature and (b) salinity for SCA.

headgroups within the micelle and consequently lowering the CMC. A similar trend has been reported for SCA (Figure 6.3b).[37]

Determining the CMC is essential for identifying the optimal surfactant concentration required to reduce IFT and alter wettability.[60] When optimizing the surfactant slug for EOR applications, it is crucial to select a concentration significantly above the CMC. This ensures that, even after accounting for surfactant losses due to adsorption and other factors under reservoir conditions, the surfactant concentration remains above the CMC. Maintaining this concentration is critical for ensuring an efficient and effective flooding process.[27]

6.3.2 Interfacial Tension Reduction

The surface and interfacial activities of surfactants are fundamental properties that determine their suitability for various applications. In EOR, surfactants primarily function to lower the IFT between crude oil and water, facilitating the release of trapped capillary oil. As a result, interfacial activity is a critical factor in selecting surfactants for cEOR.

Bio-based surfactants have exhibited IFT reduction capabilities comparable to, and in some cases exceeding, those of conventional surfactants. For example, SMES and sodium ethyl ester sulfonate (SEES) have been reported to reduce IFT to the 10^{-2} range.[26,27,29] Furthermore, bio-based zwitterionic surfactants derived from various fatty acid sources have achieved ultralow IFT levels (10^{-3}) without additives at 50°C. Notably, these surfactants are primarily dimethylcarboxylbetaines featuring a phenyl group in their hydrophobic tail.[30,32–34]

Most studies on AASs reported in the literature have yielded less satisfactory results. In these studies, kerosene was commonly used as the oleic phase, and the pendant drop method was employed for IFT measurements, making the attainment of ultralow IFT values unlikely.

These surfactants reduced IFT by only 42–55% at the CMC,[10,11,13,15,16] as illustrated in Figure 6.4a. This performance was inferior to that of conventional surfactants like SDS. However, salts of N-α-acylamino acids and fatty acid mixtures have demonstrated superior interfacial properties, comparable to those of conventional

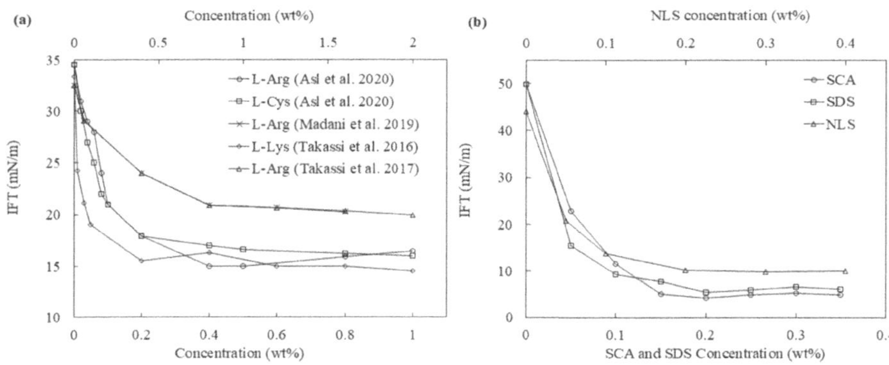

FIGURE 6.4 (a) Aqueous-kerosene IFTs for L-Arg, L-Cys, and L-Lys and (b) aqueous-decane IFTs for SCA, NLS, and SDS at 25°C.[39]

surfactants. For instance, as reported by Tackie-Otoo et al.,[39] SCA and NLS showed comparable IFT reduction performance to SDS when tested with water and n-decane, as shown in Figure 6.4b. The IFT between water and kerosene is approximately 34.5 mN/m, while the IFT between water and n-decane is around 49.82 mN/m.[37] This difference arises because kerosene is a complex mixture of hydrocarbons with carbon chains ranging from 11 to 13 atoms,[61] whereas n-decane is a single-component hydrocarbon. The ability of SCA and NLS to significantly reduce the higher IFT of water-decane to much lower values at their CMC highlights their greater potential for enhancing oil recovery compared to other AASs reported in the literature.

Deljooei et al.[44] conducted IFT studies on L-Asp with crude oil, reporting a 13.04% reduction at a concentration of 0.2 wt% (Figure 6.5a). The pendant drop method was employed for IFT measurements, although it has a lower detection limit of 0.1 mN/m.[25] Despite this limitation, the observed ~13% reduction in IFT remains relatively unsatisfactory. In contrast, the spinning drop method is capable of measuring IFT values well below 0.1 mN/m, making it more suitable for assessing ultralow IFT.

The IFT reduction capabilities of various AASs in aqueous-crude oil systems have been evaluated using a spinning drop tensiometer. According to Tackie Otoo et al.,[39] both LGA and NLS have been tested for their effectiveness in reducing IFT. Due to the presence of an additional carboxylate group in LGA's head group, which increases repulsion among surfactant molecules, NLS exhibited superior interfacial properties. As illustrated in Figures 6.5b and 6.5c, LGA achieved a minimum IFT of 0.135 mN/m at 1.2 wt%, whereas NLS reached IFT values in the 10^{-2} range at just 0.2 wt% concentration.

Similarly, Ren et al.[12] reported that diacyl amino acid sodium (DAAS) salt achieved a minimum IFT in the 10^{-2} range at a concentration of 0.3 wt%, comparable to Daqing petroleum sulfonate (DPS), as depicted in Figure 6.5d. In addition, SCA demonstrated superior interfacial properties compared to SDS, despite both surfactants exhibiting similar surface activity and aggregation behavior, as shown in Figure 6.5e.[37]

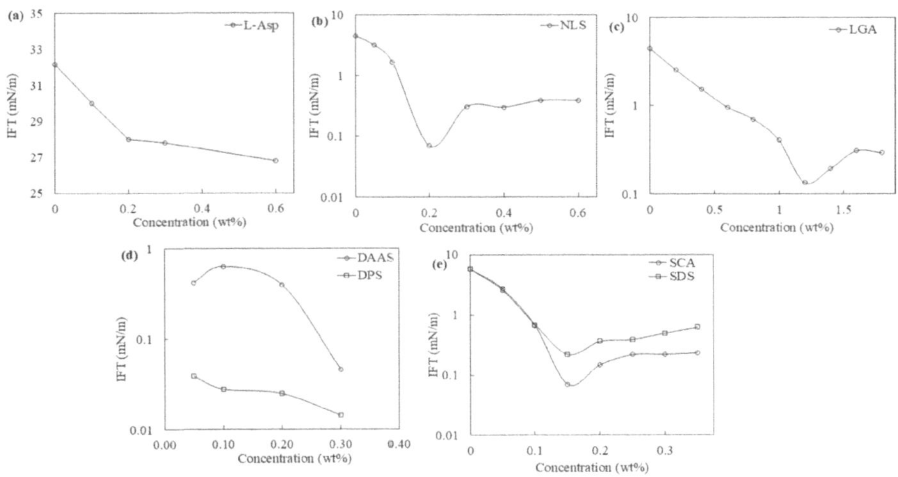

FIGURE 6.5 Aqueous-crude oil IFTs for (a) L-Asp[44], (b) NLS[39], (c) LGA[39], (d) DAAS and DPS[12], and (e) SCA and SDS at 25°C.[37]

6.3.2.1 Effect of Salinity

One of the main areas of research has been how salinity affects the ability of bio-based surfactants to reduce IFT. With their IFT reduction capabilities increasing up to a certain salinity before declining, SMES and SEES have shown good salt tolerance.[25–27,29] When the surfactant is present, this threshold is known as the ideal salinity for the aqueous-crude oil system. Strong electrolyte tolerance is another feature of bio-based zwitterionic surfactants.[31,33,34] They also interact synergistically with ions that normally cause precipitation in anionic surfactants, such as Ca^{2+} and NaCl. In these situations, IFT first falls but then increases as the electrolyte concentration rises. However, the impact of salinity varies for AASs. Rostami et al.[16] found that for L-Lys, the IFT trend increased as salinity increased, whereas Madani et al.[15] found that L-Arg was trending downward. Conversely, Deljooei et al.[44] observed that IFT fell as salinity increased until it reached an optimal level, after which it started to increase for L-Asp.

In these three studies on AASs, salinity effect is investigated by varying the surfactant concentration at different salinities. However, in most studies on bio-based surfactants, the effect of salinity is investigated at the CMC of the surfactants.[25,26,29] In these studies, synergism is observed through further reduction of the IFT in the presence of salt. Contrary to this observation, SCA, NLS and LGA tend to increase IFT in the presence of salt at their CMCs.[38,39] The presence of salt is known to enhance both the surface activity and aggregation behavior of surfactants.[22] Consequently, a decrease in the CMC is observed when salt is present. In this context, the addition of salt at the "original CMC" means that the surfactant is functioning above the CMC, which explains why the IFT is higher, as observed. Therefore, investigating the effect of salinity at a surfactant concentration lower than the CMC yielded synergistic effect like SDS as shown in Figure 6.6.

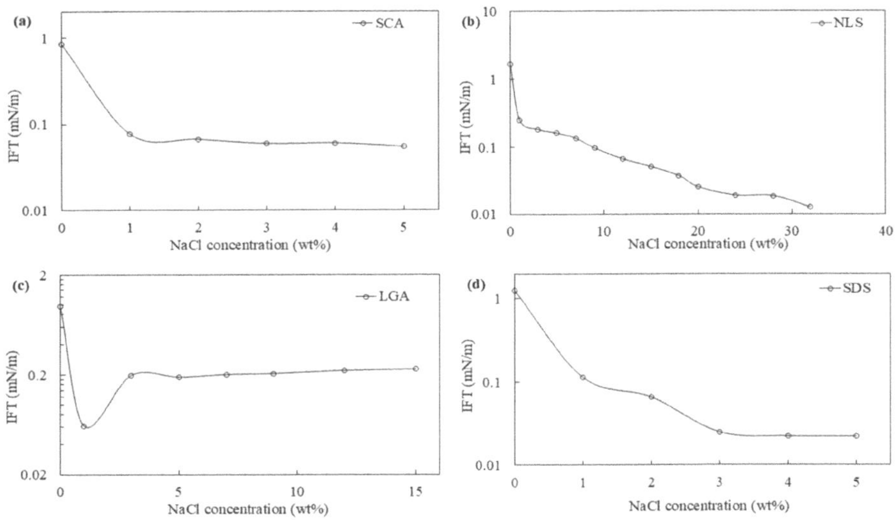

FIGURE 6.6 Salinity effect on IFT reduction by (a) SCA, (b) NLS, (c) LGA, and (d) SDS at 25°C.[38-39]

6.3.2.2 Effect of Alkalis

Alkalis are often incorporated to enhance the performance of surfactants in oil recovery processes. However, their interaction with surfactants can be either synergistic or antagonistic,[62] similar to the effects observed in the presence of salts. Alkalis react with crude oil to form in situ soap, which can combine with synthetic surfactants to create a surfactant mixture that further improves IFT reduction. In addition, alkalis function similarly to salts by promoting surfactant adsorption at the oil-water interface.[4]

The influence of alkalis, particularly the effect of increased pH on IFT reduction, has been extensively studied as part of evaluating the EOR potential of various bio-based surfactants. Saxena et al.[29] and Pal et al.[26] reported that for SEES, IFT values continuously decreased to ultralow levels as alkali concentration increased, eventually stabilizing at higher concentrations. In contrast, Saxena et al.[25] found that in the presence of alkalis such as monoethanolamine and Na_2CO_3, SMES exhibited improved IFT reduction up to a certain alkali concentration. However, beyond this threshold, a further increase in alkali concentration resulted in a rise in IFT.

For bio-based zwitterionic surfactants, carboxyl betaines have an isoelectric point of 8.2.[63] As a result, these surfactants exhibit cationic behavior below this pH and anionic behavior above it. Zhang et al.[33] reported that at a pH of 3, the resulting cationic surfactant was unable to achieve ultralow IFT values. However, when the pH was 7 or higher, the surfactant's interfacial properties remained largely unaffected.

Similarly, Xu et al.[31] provided evidence of synergistic effects when Na_2CO_3 was introduced to sulfonic zwitterionic surfactants. However, literature on the impact

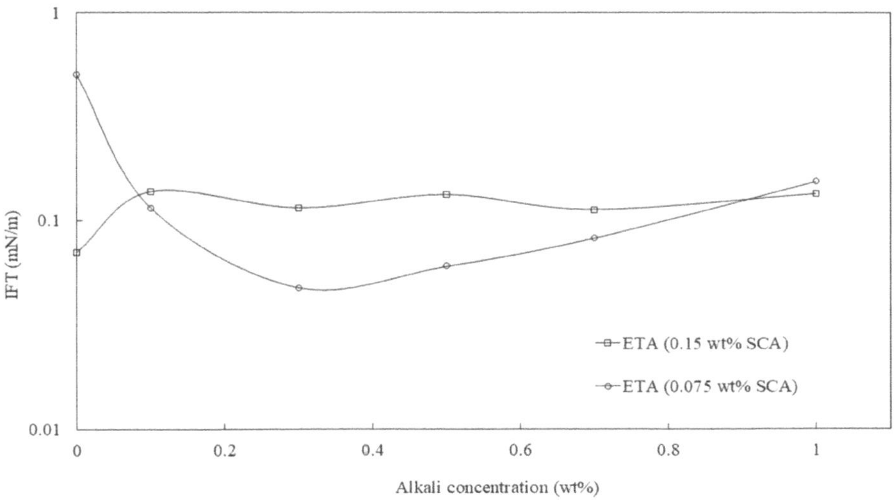

FIGURE 6.7 Effect of alkali on aqueous-crude oil IFT of SCA.[62]

of alkalis on AASs is limited, with existing studies primarily focusing on SCA. In research conducted by Tackie-Otoo et al.,[62] SCA was combined with monoethanolamine (ETA) to develop an alkali-surfactant slug for EOR. During the optimization of this formulation through IFT measurements, synergism was observed when ETA was added to SCA at a concentration lower than the CMC, as illustrated in Figure 6.7. This effect is likely due to the same mechanism described in relation to salinity effects.

6.3.2.3 Effect of Temperature

The impact of temperature on the performance of bio-based surfactants has not been extensively documented because the majority of interfacial studies have been carried out at particular temperatures of interest. However, Pal et al.[26] observed that SEES exhibited enhanced IFT reduction with increasing temperature, regardless of surfactant concentration (Figure 6.8a). This observation could be attributed to delayed micellization as observed in Figure 6.8a. The delay in micellization means there will be more surfactant molecules to adsorb onto the oil-water interface thereby reducing IFT further. SCA also exhibited improved interfacial properties with increasing temperature until 50°C after which IFT increased with increasing temperature as shown in Figure 6.8b.[38] The temperature at which the minimum IFT is achieved is referred to as the phase inversion temperature (Ye at al., 2008).[38,71] Zhang et al.[33] reported that CPDB retained strong interfacial activity even after being stored at 50°C for 100 days. Similarly, Saxena et al.[25] found that SMES maintained its interfacial properties, producing ultralow IFTs at the CMC even after being heated at 90°C for 15 days. In addition, bio-based zwitterionic surfactants have been reported to exhibit temperature insensitivity.

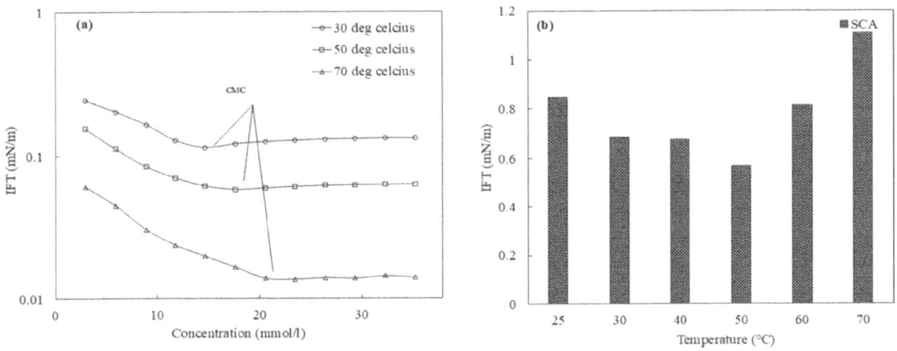

FIGURE 6.8 Temperature effect on IFT at (a) different concentration of SEES[26] and (b) 0.075 wt% concentration of SCA.[38]

6.3.3 Wettability Alteration

When fluids flow in multiple phases, the fluid that preferentially wets the rock surface loses mobility in comparison to the fluid that has a lower propensity to wet the rock. Most oil reservoirs worldwide tend to become oil-wet or intermediate-wet over time due to the prolonged storage of crude oil.[64,65] However, it is more desirable for rock surfaces to be water-wet, as this enhances the mobility of oil. Therefore, modifying the wettability of rock surfaces is a critical aspect of improving oil recovery. In EOR, one of the primary objectives of injecting chemical agents is to alter the rock surface wettability to ensure favorable mobility of reservoir fluids.

Surfactants influence rock wettability, which is typically assessed by measuring the contact angle. A contact angle below 90° signifies strong wettability, whereas an angle above 90° suggests weaker wettability of a particular phase.[66] Wettability alteration primarily occurs through two mechanisms: ion-pair formation and the interaction between surfactant molecules and crude oil components adsorbed on the rock surface.[15] In the ion-pair mechanism, anionic surfactants are particularly effective for altering the wettability of sandstone formations, while cationic surfactants are more suitable for carbonate rocks.[27]

The second mechanism involves the adsorption of the surfactant's hydrophilic group onto the rock's hydrophobic surface, facilitating wettability modification.[27] Among these processes, ion-pair formation has been found to be more effective than hydrophobic interactions in shifting rock wettability toward a more water-wet state.[67] Wettability alteration is an inherent property of surfactants, although its effectiveness varies depending on the surfactant type and rock composition. As a result, modifying reservoir rock wettability from oil-wet or intermediate-wet to a more water-wet condition is a key mechanism through which surfactants enhance oil recovery

This alteration enhances the microscopic displacement of oil.[68] Though not as predominant as IFT reduction when it comes to surfactants, wettability alteration is one of the main mechanisms investigated to test the potential of various bio-based surfactants in enhancing oil recovery. The bio-based zwitterionic surfactants have also been reported to alter the surface of a hydrophobic substrate to water-wet.[30–34]

SMES and SEES have been reported to alter rock surface wettability effectively toward water-wetness.[25–27,29] Saxana et al.[29] reported that SEES could effectively alter quartz and carbonate rock surface wettability (Figure 6.9a). Pal et al.[26] also reported that the effectiveness of SEES in altering rock surface wettability increases when the temperature rises. SMES also proved to be effective in altering quartz surface at elevated temperature as reported by Saxana et al.[25] and shown in Figure 6.9b. All studies on SMES and SEES reported the effect of salt. Like IFT reduction, the wettability alteration capabilities of SMES and SEES are improved with an increase in salt concentration to an optimal value. The effectiveness then decreases with further increase in salt concentration.[25–27,29] Saxana et al.[25] also reported that polymer addition reduces the effectiveness of wettability alteration by SMES.

As mentioned already, most AASs reported in the literature showed less satisfactory results. Though they exhibit the ability to alter rock surface wettability, they could not achieve strong water wetness. L-Lys as reported by Rostami et al.[16] could only reduce the kerosene wet carbonate surface (strong oil wet) to weak oil wet (Figure 6.10a). Asl et al.[13] reported an improved performance of L-Arg and L-Cys on kerosene wet carbonate surface. Both surfactants altered the carbonate surface from strong oil wet to weak water wet (Figure 6.10b). Madani et al.[15] on the other hand,

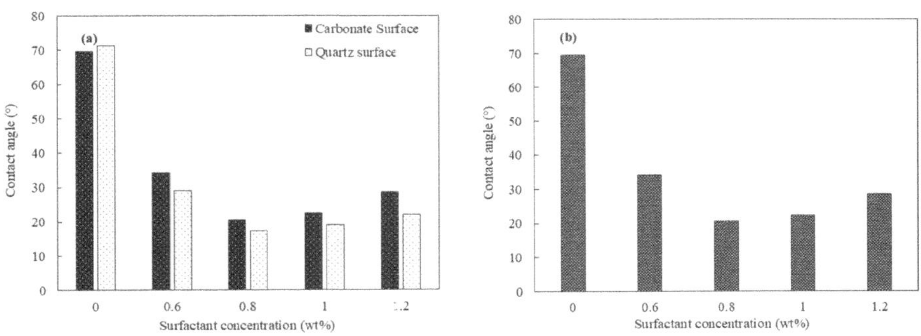

FIGURE 6.9 Final contact angle after 1000 s on (a) carbonate and quartz surface for SEES at varying concentrations at 27°C[29] and (b) quartz surface for SMES at varying concentrations at 75°C.[25]

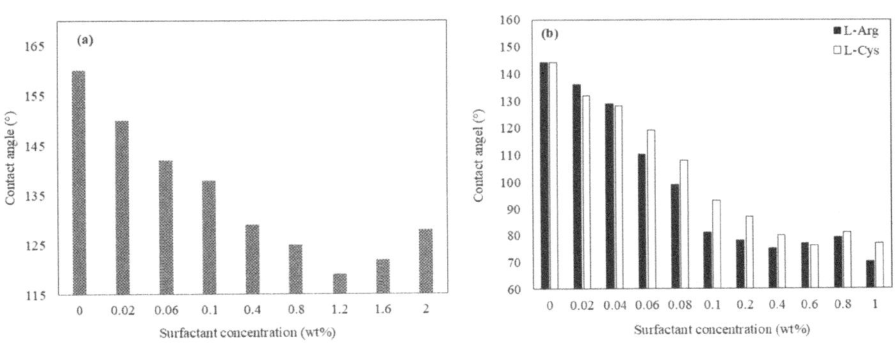

FIGURE 6.10 Contact angle variation with concentration on kerosene wet carbonate surface for (a) L-Lys[16] and (b) L-Arg and L-Cys.[13]

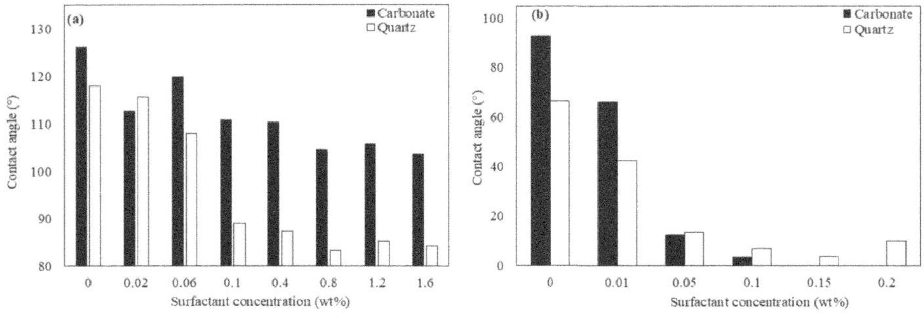

FIGURE 6.11 Contact angle variation with concentration for (a) L-Arg on kerosene wet carbonate and quartz surface[15] and (b) SCA on oil wet carbonate and quartz surface.[37]

reported the performance of L-Arg on both kerosene wet carbonate and sandstone surface. The carbonate surface was only rendered poor oil wet while the sandstone was altered from oil wet to weak water wet (Figure 6.11a). SCA on the other hand has been reported to be effective in altering the wettability of quartz and carbonate rock surfaces.[37,38] The quartz surface was altered from intermediate wet to strong water wetness while the carbonate surface was altered from oil wet to strong water wet (Figure 6.11b). NLS and LGA also altered intermediate wet quartz surface to strong water wet.[39]

6.3.4 Adsorption

As previously discussed, surfactant adsorption plays a crucial role in reducing IFT and altering the wettability of rocks, both essential mechanisms for effective surfactant flooding. However, excessive surfactant adsorption can negatively impact the process by increasing the surfactant concentration required for efficient oil recovery, thus affecting project economics. Therefore, understanding the adsorption characteristics of surfactants is critical in cEOR to optimize slug design and improve the cost-effectiveness of the process.

There is currently little literature on the adsorption behavior of bio-based surfactants. Saxena et al.[69] examined J-SMES adsorption in 2018, and it was discovered that SMES adsorption closely resembles the Langmuir isotherm model, suggesting a successful adsorption process onto sand surfaces. In addition, their analysis showed that the pseudo-second-order kinetic model best describes the adsorption kinetics. In addition, the study showed that because the process is exothermic, adsorption density decreases at higher temperatures. The majority of research on AAS's EOR potential has not concentrated on their adsorption properties. Tackie-Otoo et al.[62] investigated how SCA adsorbs onto sand and how alkali concentration affects this behavior (Figure 6.12a). SCA showed Langmuir-type behavior, which is indicative of favorable adsorption, like the majority of surfactants. However, SCA adsorption exhibited Freundlich behavior at an alkali concentration of 0.3 wt%, rendering adsorption less favourable. There were negligible decreases in SCA adsorption above this alkali concentration (Figure 6.12b). To determine whether AASs and bio-based zwitterionic

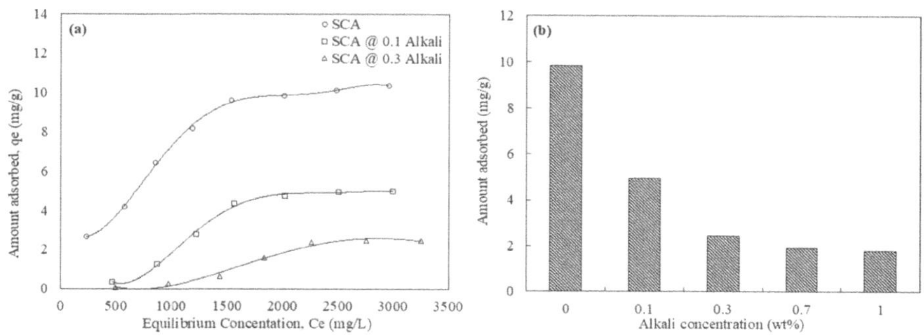

FIGURE 6.12 (a) Adsorption isotherm of SCA and (b) alkali effect on SCA adsorption on the quartz surface.[62]

surfactants are appropriate for a given reservoir condition, more research on their adsorption properties in diverse rock types is required.

6.3.5 Emulsification

In surfactant flooding, emulsification and entrainment serve as the primary mechanisms for transporting oil droplets. The ability of surfactants to emulsify oil by reducing IFT plays a crucial role in enhancing oil recovery. Depending on the surfactant flooding process, two immiscible fluids combined with an emulsifying agent can form a variety of emulsions. Macroemulsions, commonly referred to as emulsions, are formed at low surfactant concentrations and require vigorous agitation for their generation.[68] While these emulsions exhibit kinetic stability, they are not thermodynamically stable.

Emulsification studies primarily assess the stability of the formed emulsions by analyzing phase volume ratio variations and droplet size distribution.[22] In addition to macroemulsions, microemulsions play a vital role in surfactant flooding, particularly in micellar flooding, which involves higher surfactant concentrations.[68] These microemulsions are composed of an optimal mixture of surfactant, cosurfactant, water, and oil. Unlike macroemulsions, they are thermodynamically stable and can form spontaneously or with minimal mixing.[22]

Microemulsions are particularly advantageous for oil recovery due to their ability to achieve ultra-low IFT at the microemulsion-crude oil interface. This property is crucial for effectively displacing residual oil trapped within rock capillaries. The ability of bio-based surfactants to emulsify and the stability of the resulting emulsions have been assessed. Saxena et al.[43] looked into the stability of SMES emulsions and found that emulsions formed at surfactant concentrations close to the CMC did not phase separate for more than 180 min. In addition, emulsions formed with longer-chain oils exhibited greater stability compared to those with shorter-chain oils (Figure 6.13).[43]

Similarly, Pal et al.[26] studied the emulsion stability of SEES. After homogenizing a crude oil-surfactant mixture at the CMC for 40 min, the resulting emulsion remained stable for 21 days without phase separation at 70°C (Figure 6.14).

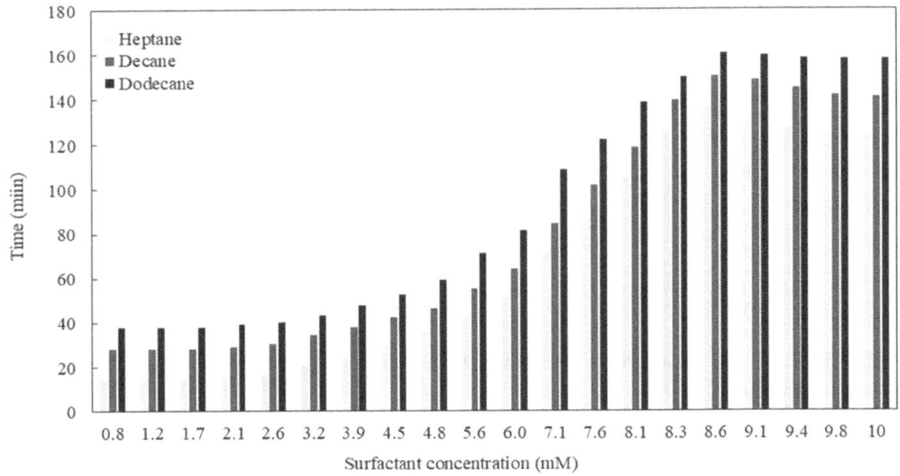

FIGURE 6.13 Emulsion stability test for SMES at various concentrations.[43]

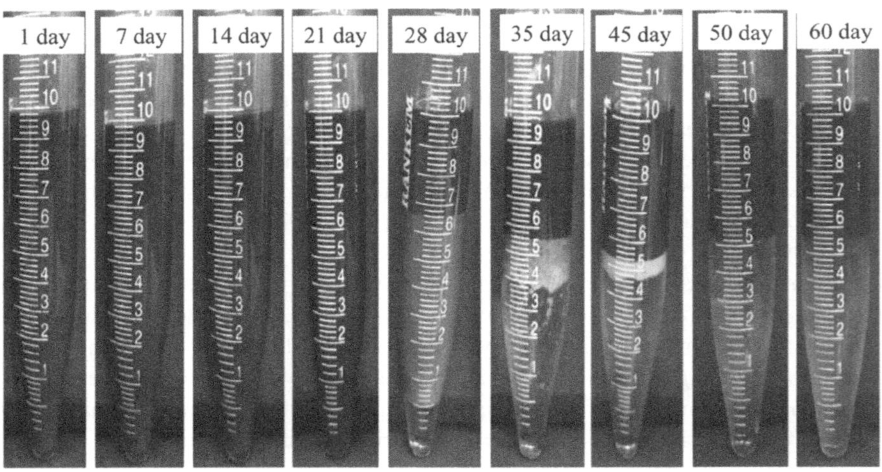

FIGURE 6.14 Stable emulsion formed in the presence of SEES at the CMC at 70°C.[26]

The emulsification properties of bio-based zwitterionic surfactants have also been assessed by measuring the time required for the separation of 3 and 5 mL aqueous phases. Zhang et al.[34] reported separation times of 170 and 302 s for WPDB at 3 and 5 mL, respectively. Similarly, Xu et al.[31] recorded separation times of 283.6 and 504.4 s for BSZS, further demonstrating its strong emulsification capability. For AASs, SCA, NLS, and LGA have proven to be highly effective emulsifying agents.

As reported by Tackie-Otoo et al.,[38] SCA forms stable emulsions, especially at very low surfactant concentration (0.05 wt%) at both room temperature and high temperature (80°C) for 1-week observation (Figure 6.15). NLS and LGA could maintain emulsion formed for a 1-week duration at 80°C without full disintegration of the emulsions formed (Figure 6.16) as reported by Tackie-Otoo et al.[39]

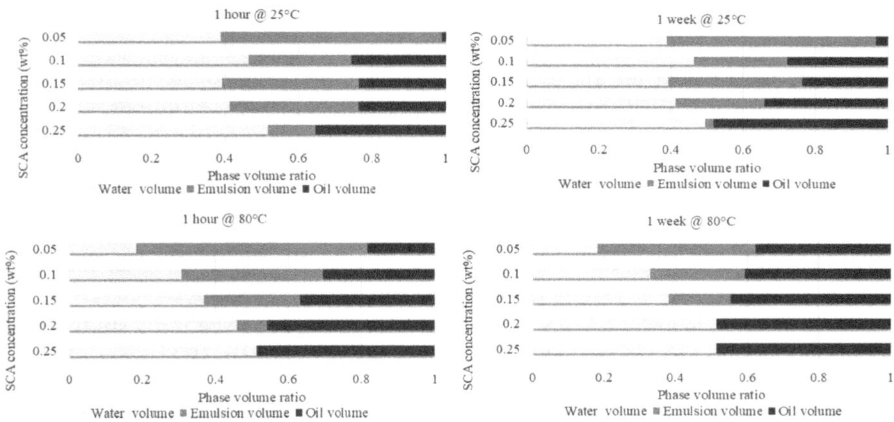

FIGURE 6.15 Stability test of emulsions formed by SCA at the various concentrations at 25°C and 80°C.[38]

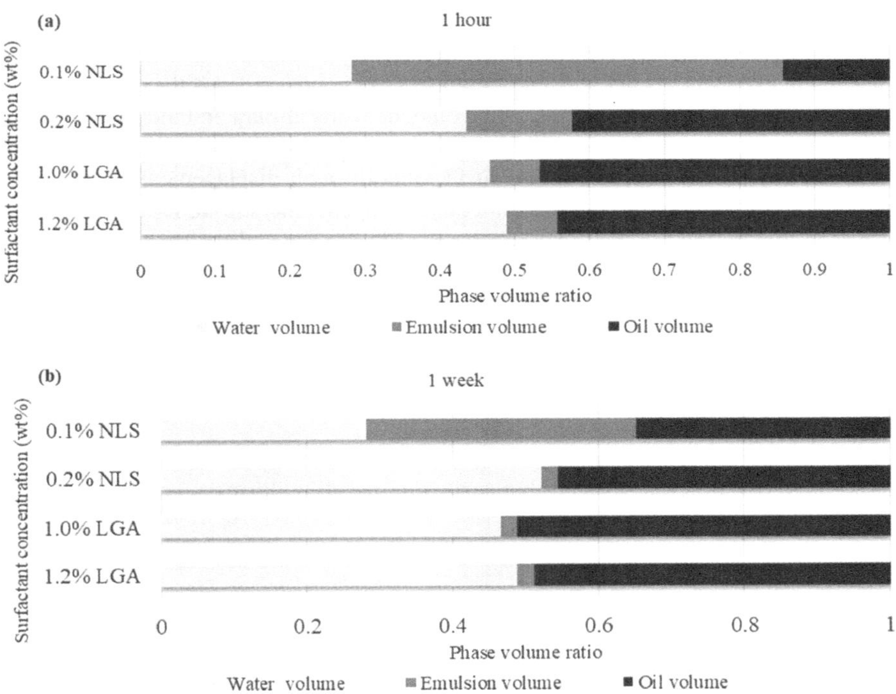

FIGURE 6.16 Stability of emulsions formed by NLS and LGA at various concentrations at 80°C.[39]

Microemulsion studies are crucial for determining the optimal salinity required to achieve key properties such as ultralow IFT, thermodynamic stability, and efficient solubilization of oil and water. Pal et al.[23,24] investigated the microemulsion system formed by SMES in the presence of a cosurfactant. Their findings confirmed SMES's ability to form microemulsions, showing that increasing the cosurfactant (propan-2-ol) enhances both the spontaneity of microemulsion formation and its stability. In addition, their studies with different single-component oils demonstrated that increasing the alkyl chain length of the oil expands the three-phase microemulsion region (Winsor type III), suggesting that heavier oils require a higher surfactant concentration for a more extensive three-phase microemulsion.

The phase behavior of SMES in microemulsions was analyzed in a study, which identified an optimal salinity of 2 wt% NaCl, where ultralow IFTs were observed across various interfaces involving microemulsion, water, and crude oil, as determined using Chun Huh's equation.[25]

6.3.6 Oil recovery Potential

In evaluating the effectiveness of a surfactant in cEOR, all previously discussed phenomena contribute to the recovery process and explain the results observed when surfactants are used. To determine the true recovery potential of a surfactant, oil displacement experiments in cores and sandpacks are conducted. All these experiments combined help to understand the various mechanisms at play and aid in optimizing the surfactant formulation for field applications. Numerous studies have examined the recovery potential of bio-based surfactants through displacement experiments. Almost all EOR research on different chemical agents includes assessing the recovery potential. Kumar et al.[27] and Saxena et al.[29] investigated the recovery potential of SMES and SEES at various concentrations. Their studies indicated that higher concentrations of these surfactants resulted in greater oil recovery (Figure 6.17). This is because the increased surfactant concentration effectively reduces IFT and alters wettability, compensating for any surfactant losses that occur during the process. No core displacement studies have been conducted for bio-based zwitterionic surfactants.

According to research investigating the possible uses of AASs, Rostami et al.[16] as well as Madani et al.[15] found that surfactant flooding during secondary recovery

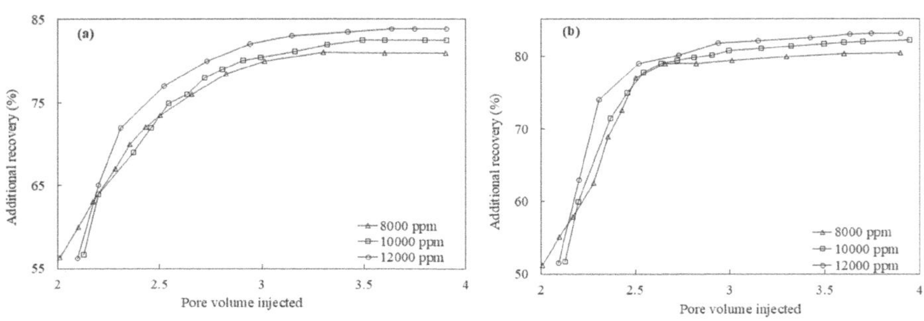

FIGURE 6.17 Additional recovery by (a) SMES[27] and (b) SEES[29] at different concentration.

produced a greater oil recovery than both tertiary surfactant flooding and secondary water flooding. Water saturation was reduced, which was the reason for this improvement. It is still necessary to assess the secondary surfactant flooding's economic viability. Asl et al.[13] also found that L-Cys and L-Arg exhibited similar performance in both IFT reduction and wettability alteration. Despite this, L-Arg demonstrated greater efficiency in displacement experiments compared to L-Cys (Figure 6.18a). This difference was attributed to higher adsorption of L-Cys on the rock surface, although no adsorption studies were conducted to confirm this hypothesis.

For NLS and LGA, similar trends were observed in their IFT reduction and wettability alteration performances, with NLS displaying superior oil recovery potential compared to LGA (Figure 6.18b), as reported by Tackie-Otoo et al.[39]

For these bio-based surfactants to be considered viable alternatives to conventional surfactants, they must demonstrate comparable or superior performance. Such comparative study has been reported by Tackie-Otoo et al.[38] Authors reported SCA to exhibit superior oil recovery potential to SDS (Figure 6.19). This observation

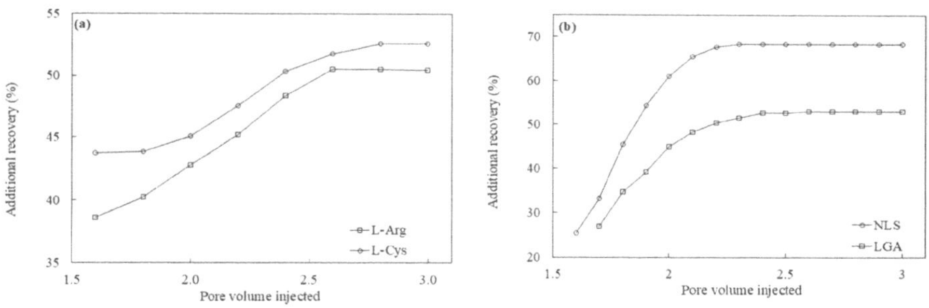

FIGURE 6.18 Additional recovery by (a) L-Arg and L-Cys[13] and (b) NLS and LGA.[39]

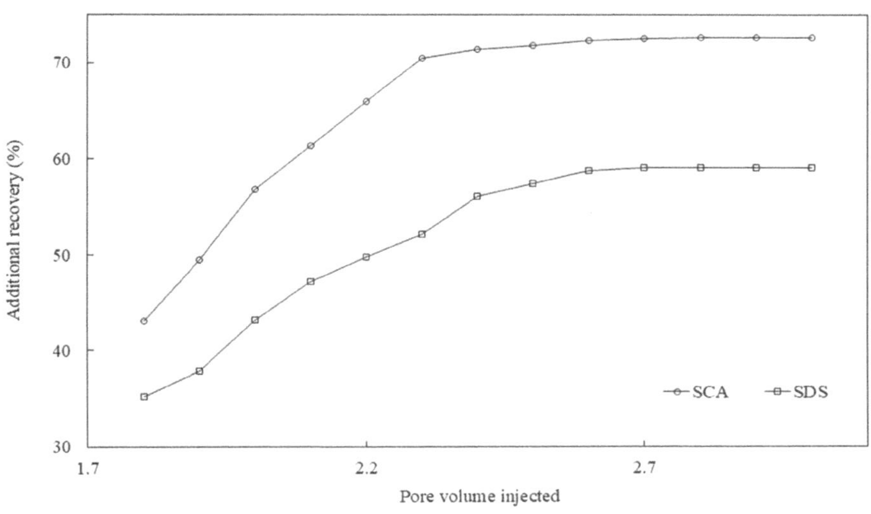

FIGURE 6.19 Additional recovery by SCA and SDS.[38]

TABLE 6.5
Oil Recovery Performance of Various AASs

Surfactant	Core Type	Brine Salinity (ppm)	Additional Oil Recovery (%OOIP)	Reference
L-Lys	Carbonate	50,000	10	16
L-Arg	Carbonate	50,000	2.49	15
L-Arg	Carbonate	100,000	11.9	13
L-Cys	Carbonate	100,000	8.9	13
L-Asp	Sandstone	20,000	11.19	44
		40,000	6.04	
SCA	Sandstone	14,000	29.53	38
NLS	Sandstone	14,000	43.06	39
LGA	Sandstone	14,000	24.56	39

corroborates the performance of SCA compared to SDS in reducing IFT and altering rock surface wettability.[37,38] The oil recovery performance of various AASs is summarized in Table 6.4. NLS, SCA and LGA showed very promising performance compared to the other AASs reported in literature. Nevertheless, a better comparison could be made when these chemicals are tested under the same conditions. NLS, SCA, and LGA are commercially available AASs mostly used in other industries while the others were newly synthesized AASs intended for EOR applications. The properties of these commercially available AASs should be studied thoroughly and considered in the synthesis of the novel ones to ensure promising oil recovery potential.

Researchers have tried to use the surfactant's ability to change wettability, rheological characteristics, and other elements known to affect the recovery process to explain the observed oil recovery in all recovery experiments. However, core and sandpack flooding alone cannot provide a complete understanding of the precise mechanisms underlying the observed recovery. To better understand displacement dynamics and determine which presumptive mechanisms are affecting recovery, a visualization model is necessary. Because they allow for pore-scale observation and analysis of the flooding process, micromodels are frequently employed for this purpose. Contritely, the majority of bio-based surfactants that have been researched in the literature have not had their displacement efficiency assessed at the pore level. Only Deljooei et al.[44] have used a microfluidic model to study the recovery potential of L-Asp. By contrasting the performance of different formulations in wettability alteration studies with the recovery attained in the micromodel experiment, they deduced that wettability alteration was their main recovery mechanism. However, the micromodel experiment did not involve pore visualization. Thus, it is strongly advised that more research be done on the recovery mechanisms of bio-based surfactants, especially at the pore-scale level.

6.4 AMINO ACIDS AS EOR AGENTS

The above discussions have been on the EOR potential of AASs surfactant among other bio-based surfactants that have been proposed and investigated in literature.

Amino acids among other natural surfactants have also been investigated for their potential to act as EOR agents. However, there are limited studies in the literature on amino acids as mentioned earlier. These investigations included testing the potential of amino acids as wettability modifier and their ability to reduce IFT. The findings of these studies are discussed in this section.

6.4.1 Interfacial Tension Reduction

The motivation to propose amino acids for reducing oil-water IFT originated from their amphipathic nature.[17] Atta et al.[21] tested various amino acids including L-arginine, L-lysine, L-tryptophan, and L-methionine for their ability to reduce decane-water and crude oil-water IFT. The amino acids did not exhibit promising ability in reducing decane-water IFT compared to conventional surfactants (SDS and CTAB); however, the L-arginine was better than the other three (Figure 6.20a). Though no reason was given for this observation in their study, this performance corroborates the report of Hung et al.[17] They noted that the amphipathic nature of arginine enables proteins to interact closely with both polar and nonpolar surface ligands simultaneously.[17] Nevertheless, L-arginine and L-lysine showed promising potential in reducing crude oil-water IFT compared to SDS and CTAB (Figure 6.20b). Atta et al.[21] attributed this observation to the basic nature of L-arginine and L-lysine. They proposed the possibility of in situ saponification occurring as in alkaline flooding.

6.4.2 Wettability Alteration

Amino acids have also been investigated as additives to brine to improve waterflooding in carbonate reservoirs. In this case, the main EOR mechanism is wettability alteration. As explained earlier, a primary mechanism in wettability alteration is the ionic interaction. An essential property of amino acids in terms of their ability to alter rock surface wettability is that their overall charge depends on the pH of the solution.[20] The pH at which an amino acid is electrically neutral is known as the isoelectric point (pI). When the solution pH is above the pI, the amino acid carries a net

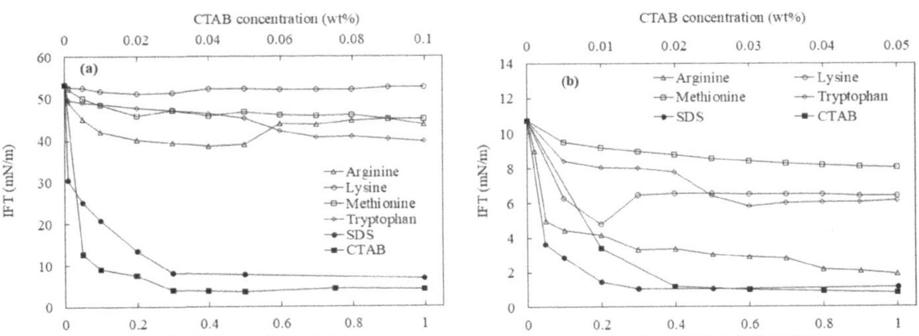

FIGURE 6.20 (a) Decane-aqueous IFT and (b) crude oil-aqueous IFT versus concentration for various amino acids, SDS, and CTAB.[21]

negative charge, whereas when the pH is below the pI, it carries a net positive charge. This change in charge occurs due to the deprotonation (release of H^+ ions) from the carboxyl and ammonium groups as the pH increases.[70]

Orozco et al.[20] investigated the wettability alteration capabilities of two amino acids, glycine and β-alanine, on polished calcite surfaces. They conducted two types of contact angle experiments: one in deionized water at room temperature and another in brine at elevated temperature. In the first case, authors reported that glycine can transform the wettability of oil-aged calcite from being oil-wet to highly water-wet when the pH of the solution is within the range between the pI of glycine and the point of zero charge (pzc) of calcite. Wettability alteration was also observed at pH values below the pI of glycine, which was attributed to the dissolution of calcite. The β-alanine showed similar results, but the contact angles were higher which was attributed to high pI leading to smaller pH range between the pI of alanine and pzc if calcite. In brine at elevated temperature, glycine was still effective in modifying the wettability of the polished calcite surface while β-alanine was not. Their imbibition experiment confirmed the ability of glycine to improve oil recovery through wettability alteration (Figure 6.21).

On the other hand, Atta et al.[21] investigated the wettability alteration capabilities of four amino acids on polished sandstone surfaces. They reported that all the amino acids studied demonstrated the ability to change the sandstone surface from oil-wet to water-wet. However, L-arginine was the most effective, followed by L-lysine, L-tryptophan, and L-methionine, as shown in Figure 6.22. For all the amino acids tested, wettability alteration was more effective at lower pH levels. This was

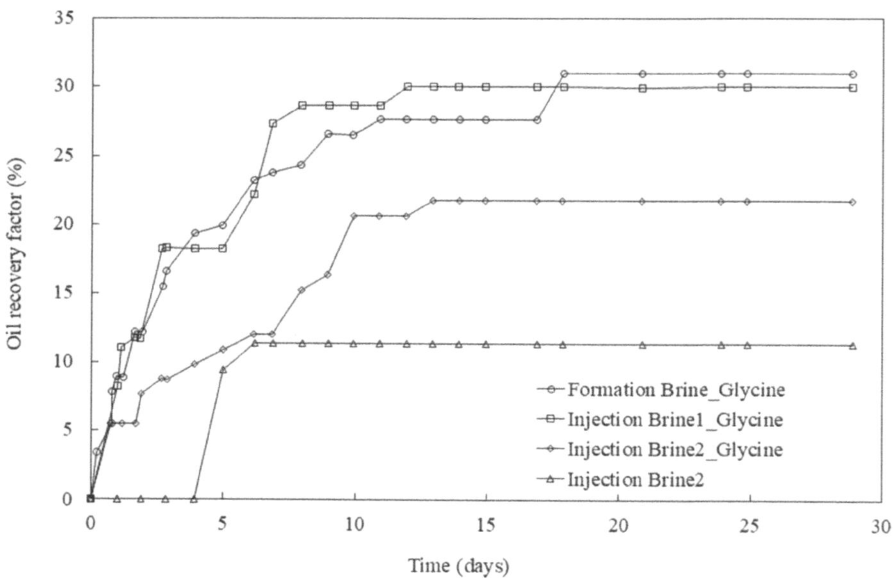

FIGURE 6.21 Oil recovery through spontaneous imbibition by combination of different brines and glycine at 90°C.[20]

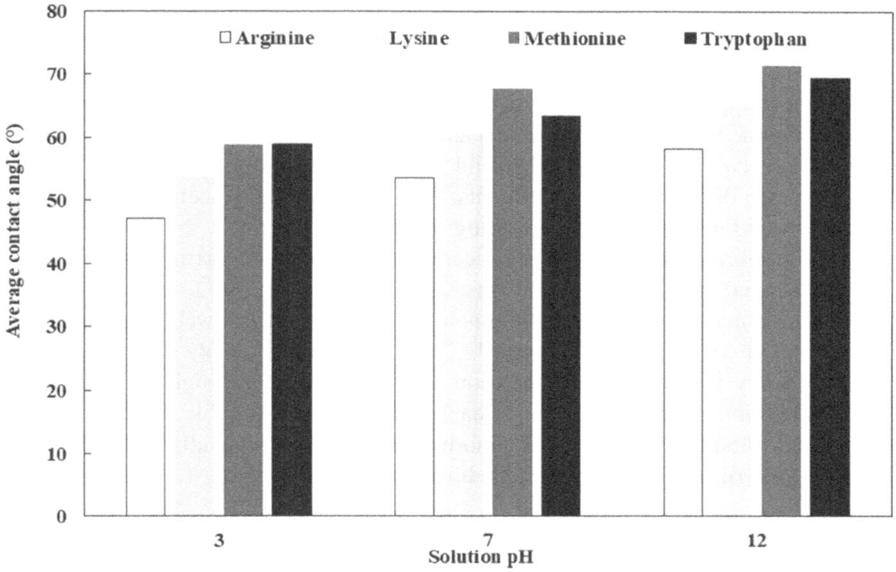

FIGURE 6.22 Wettability alteration capability of different amino acids at various pH.[21]

attributed to the pH of the solution falling within the range between the pzc of the rock surface and the pI of the amino acids. Atta et al.[21] explained that within this pH range, the sandstone surface and the amino acid carry opposite charges, enabling electrostatic attraction. The greater the difference between the mineral's pzc and the amino acids pI, the more significant the charge disparity, leading to more effective wettability alteration.

6.5 CONCLUSIONS

Amino acids are useful in many industrial applications, and this has drawn attention to research their applicability in other areas like the oil and gas industry. In this chapter, the potential of amino acids as chemical agents for EOR or as additives to improve waterflooding has been discussed. Although amino acids tend to exhibit amphiphilic behavior, they lack long alkyl chain length which limits their effectiveness in EOR application. However, deprivation of amino acids or namely AASs among other bio-based surfactants has been investigated and proven to have promising potential to be deployed for EOR. Based on the investigations discussed, the following conclusions can be drawn:

- SMES and SEES exhibit promising potential in reducing oil-water IFT and altering the wettability of both quartz and carbonate surface. Bio-based zwitterionic surfactants have excellent interfacial properties achieving ultralow IFT without additives and altering hydrophobic surface to more hydrophilic. Most of the reported AASs showed less promising EOR potential, especially in IFT reduction and wettability alteration. Nevertheless,

commercially available AASs like SCA, NLS, and LGA showed promising IFT reduction and wettability alteration capabilities.
- The emulsification studies on these surfactants corroborated their ability to reduce oi_L-water IFT. Adsorption studies though limited are also included which helps to optimize the surfactant formulation to ensure effective and efficient performance. SMES and SEES proved to be suitable for EOR through displacement experiments. All AASs that have been reported also showed the potential of enhancing oil recovery, but SCA, NLS, and LGA were more promising. No displacement experiment report is available for bio-based zwitterionic surfactants.
- Amino acids showed promising potential in rock surface wettability alteration but were less promising in IFT reduction. Their ability to improve oil recovery via wettability alteration has been proven through spontaneous imbibition experiments. This means they are more suitable for improving waterflooding. They serve as a more environmentally friendly and economical alternative to conventional additives.

These alternative chemical agents for improving oil recovery lack deployment in the field. Their reports are limited to laboratory investigations. Not even a pilot test has been reported on their potential. The main limitation is the economic feasibility of their deployment. The viability of deploying these chemicals depends on their production in commercial quantities. AASs like SCA, NLS, and LGA are available commercially, however, not in the quantity suitable for EOR application as they are mostly deployed in the pharmaceuticals and cosmetics industry. Thorough economic investigations are required to ensure the deployment of these alternative chemical agents in field applications.

ACKNOWLEDGMENTS

The authors gratefully acknowledge the support of University Teknologi Petronas and the Centre of Research in Enhanced Oil Recovery.

REFERENCES

1. Negin C, Ali S, Xie Q. Most common surfactants employed in chemical enhanced oil recovery. *Petroleum*. 2017;3(2):197–211. doi: 10.1016/j.petlm.2016.11.007
2. Mandal A. Chemical flood enhanced oil recovery: a review. *Int. J. Oil, Gas Coal Technol.* 2015;9(3):241. doi: 10.1504/IJOGCT.2015.069001
3. Gbadamosi AO, Junin R, Manan MA, Agi A, Yusuff AS. An overview of chemical enhanced oil recovery: recent advances and prospects. *Int Nano*. 2019;9:171–202.
4. Olajire AA. Review of ASP EOR (alkaline surfactant polymer enhanced oil recovery) technology in the petroleum industry : prospects and challenges. *Energy* [Internet]. 2014;77:963–982. doi: 10.1016/j.energy.2014.09.005
5. Nakama Y. Chapter 15- Surfactants. In: Sakamoto K, Lochhead RY, Maibach HI, Yamashita YBT- CS and T, editors. *Cosmetic science and technology: theoretical principles and applications*. Amsterdam: Elsevier; 2017. p. 231–244. doi:https://doi.org/10.1016/B978-0-12-802005-0.00015-X

6. Morán MC, Pinazo A, Pérez L, Clapés P, Angelet M, García MT, Vinardell MP, Infante MR. "Green" amino acid-based surfactants. *Green Chem.* 2004;6(5):233–240.
7. Chandra N, Tyagi VK. Synthesis, properties, and applications of amino acids based surfactants: a review. *J Dispers Sci Technol.* 2013;34(6):800–808.
8. Atta DY, Negash BM, Yekeen N, Habte AD. A state-of-the-art review on the application of natural surfactants in enhanced oil recovery. *J Mol Liq.* 2020;321:114888. doi: 10.1016/j.molliq.2020.114888
9. Tackie-Otoo BN, Ayoub Mohammed MA, Yekeen N, Negash BM. Alternative chemical agents for alkalis, surfactants and polymers for enhanced oil recovery: research trend and prospects. *J Pet Sci Eng.* 2020;187:1–32. doi: 10.1016/j.petrol.2019.106828
10. Takassi MA, Hashemi A, Rostami A. A lysine amino acid-based surfactant : application in enhanced oil recovery a lysine amino acid-based surfactant : application in enhanced oil recovery. *Pet Sci Technol.* 2016;34(17–18):1521–1526. doi: 10.1080/10916466.2016.1205605
11. Takassi MA, Zargar G, Madani M, Zadehnazari A. The preparation of an amino acid-based surfactant and its potential application as an EOR agent. *Pet Sci Technol.* 2017;35(4):385–391. doi: 10.1080/10916466.2016.1238933
12. Ren H, Shi C, Song S. Synthesis of diacyl amino acid surfactant and evaluation of its potential for surfactant – polymer flooding. *Appl Petrochem Res.* 2016;6(1):59–63. doi: 10.1007/s13203-014-0094–6
13. Asl HF, Zargar G, Manshad AK, Takassi MA, Ali JA, Keshavarz A. Experimental investigation into l-Arg and l-Cys eco-friendly surfactants in enhanced oil recovery by considering IFT reduction and wettability alteration. *Pet Sci.* 2020;17(1):105–117.
14. Asl HF, Zargar G, Manshad AK, Takassi MA, Ali JA, Keshavarz A. Effect of SiO2 nanoparticles on the performance of L-Arg and L-Cys surfactants for enhanced oil recovery in carbonate porous media. *J Mol Liq.* 2020;300:112290.
15. Madani M, Zargar G, Takassi MA, Daryasafar A, Wood DA, Zhang Z. Fundamental investigation of an environmentally-friendly surfactant agent for chemical enhanced oil recovery. *Fuel.* 2019;238:186–197.
16. Rostami A, Hashemi A, Takassi MA, Zadehnazari A. Experimental assessment of a lysine derivative surfactant for enhanced oil recovery in carbonate rocks: mechanistic and core displacement analysis. *J Mol Liq [Internet].* 2017;232(October):310–318. doi: 10.1016/j.molliq.2017.01.042
17. Hung A, Mager M, Hembury M, Stellacci F, Stevens MM, Yarovsky I. Amphiphilic amino acids: a key to adsorbing proteins to nanopatterned surfaces? *Chem Sci.* 2013;4(3):928–937.
18. Watry MR, Richmond GL. Orientation and conformation of amino acids in monolayers adsorbed at an oil/water interface as determined by vibrational sum-frequency spectroscopy. *J Phys Chem B.* 2002;106(48):12517–12523.
19. Ghosh N, Dutta P, Das KP, Chattoraj DK. Thermodynamics of adsorption of L-amino acids at oil–water interfaces. *J Solution Chem.* 2003;32:1045–1064.
20. Lara Orozco RA, Abeykoon GA, Wang M, Argüelles-Vivas F, Okuno R, Lake LW, Ayirala SC, AlSofi AM. Amino acid as a novel wettability modifier for enhanced waterflooding in carbonate reservoirs. *SPE Reserv. Eval. Eng.* 2020;23(02):741–757.
21. Atta DY, Negash BM, Yekeen N, Habte AD, Malik ABA. Influence of natural L-amino acids on the interfacial tension of an oil-water system and rock wettability alterations. *J Pet Sci Eng.* 2020;199:108241. doi: https://doi.org/10.1016/j.petrol.2020.108241
22. Rosen MJ, Kunjappu JT. *Surfactants and interfacial phenomena.* [place unknown]: John Wiley & Sons; 2012.
23. Pal N, Saxena N, Mandal A. Phase behavior, solubilization, and phase transition of a microemulsion system stabilized by a novel surfactant synthesized from castor oil. *J Chem Eng Data.* 2017;62(4):1278–1291.

24. Pal N, Kumar S, Bera A, Mandal A. Phase behaviour and characterization of microemulsion stabilized by a novel synthesized surfactant: implications for enhanced oil recovery. *Fuel*. 2019;235:995–1009.
25. Saxena N, Goswami A, Dhodapkar PK, Nihalani MC, Mandal A. Bio-based surfactant for enhanced oil recovery: interfacial properties, emulsification and rock-fluid interactions. *J Pet Sci Eng*. 2019;176:299–311.
26. Pal N, Saxena N, Divya Laxmi KV, Mandal A. Interfacial behaviour, wettability alteration and emulsification characteristics of a novel surfactant: implications for enhanced oil recovery. *Chem Eng Sci*. 2018;187:200–212. doi: 10.1016/j.ces.2018.04.062
27. Kumar S, Kumar A, Mandal A. Characterizations of surfactant synthesized from Jatropha oil and its application in enhanced oil recovery. *AIChE Journal*. 2017;63(7):2731–2741.
28. Kumar A, Mandal A. Synthesis and physiochemical characterization of zwitterionic surfactant for application in enhanced oil recovery. *J Mol Liq [Internet]*. 2017;243:61–71. doi: 10.1016/j.molliq.2017.08.032
29. Saxena N, Pal N, Dey S, Mandal A. Characterizations of surfactant synthesized from palm oil and its application in enhanced oil recovery. *J Taiwan Inst Chem Eng*. 2017;81: 343–355.
30. Zhang Q-Q, Cai B-X, Gang H-Z, Yang S-Z, Mu B-Z. A family of novel bio-based zwitterionic surfactants derived from oleic acid. *RSC Adv*. 2014;4(72):38393–38396.
31. Xu W, Bian P, Gang H, Liu J, Mu B, Yang S. A novel bio-based sulfonic zwitterionic surfactant derived from transgenic soybean oil and its performance in surface and interfacial activities. *J. Pet. Sci. Technol*. 2018;8(1):32–44.
32. Gang H, Zhang Q, Wang W, Cai B, Liu J, Yang S, Mu B. Synthesis and interfacial properties of bio-based zwitterionic surfactants derived from different fatty acids in non-edible vegetable oils. *J Renew Mater*. 2020;8(4):417.
33. Zhang QQ, Cai BX, Xu WJ, Gang HZ, Liu JF, Yang SZ, Mu BZ. Novel zwitterionic surfactant derived from castor oil and its performance evaluation for oil recovery. *Colloids Surf A Physicochem Eng Asp [Internet]*. 2015;483:87–95. doi: 10.1016/j.colsurfa.2015.05.060
34. Zhang QQ, Cai B-X, Xu W-J, Gang H-Z, Liu J-F, Yang S-Z, Mu B-Z. The rebirth of waste cooking oil to novel bio-based surfactants. *Sci Rep*. 2015;5(1):1–7.
35. Zhang G, Xu B, Han F, Zhou Y, Liu H, Li Y, Cui L, Tan T, Wang N. Green synthesis, composition analysis and surface active properties of sodium cocoyl glycinate. *Am J Analyt Chem*. 2013;4:445–450.
36. Rondel C, Alric I, Mouloungui Z, Blanco J, Silvestre F. Synthesis and properties of lipoamino acid–fatty acid mixtures: influence of the amphiphilic structure. *J Surfactants Deterg*. 2009;12(3):269–275.
37. Tackie-Otoo BN, Mohammed MAA. Experimental investigation of the behaviour of a novel amino acid-based surfactant relevant to EOR application. *J Mol Liq [Internet]*. 2020;316:113848. doi: https://doi.org/10.1016/j.molliq.2020.113848
38. Tackie-Otoo BN, Ayoub Mohammed MA, Owusu EB. Investigation of the enhanced oil recovery potential of sodium cocoyl alaninate: an eco-friendly surfactant. *J Pet Explor Prod Technol*. 2022;12:2785–2799.
39. Tackie-Otoo BN, Mohammed MAA, Tze JYS, Hassan AM. Experimental investigation of N-lauroyl sarcosine and N-lauroyl-L-glutamic acid as green surfactants for enhanced oil recovery application. *J Mol Liq*. 2022;362:119738.
40. Elraies KA, Tan IM, Awang M, Saaid I. The synthesis and performance of sodium methyl ester sulfonate for enhanced oil recovery. *Pet Sci Technol*. 2010;28(17):1799–1806.
41. Babu K, Maurya NK, Mandal A, Saxena VK. Synthesis and characterization of sodium methyl ester sulfonate for chemically-enhanced oil recovery. *Braz J Chem Eng*. 2015;32:795–803.

42. Babu K, Pal N, Bera A, Saxena VK, Mandal A. Studies on interfacial tension and contact angle of synthesized surfactant and polymeric from castor oil for enhanced oil recovery. *Appl Surf Sci.* 2015;353:1126–1136.
43. Saxena N, Pal N, Ojha K, Dey S, Mandal A. Synthesis, characterization, physical and thermodynamic properties of a novel anionic surfactant derived from Sapindus laurifolius. *RSC Adv.* 2018;8(43):24485–24499.
44. Deljooei M, Zargar G, Nooripoor V, Takassi MA, Esfandiarian A. Novel green surfactant made from L-aspartic acid as enhancer of oil production from sandstone reservoirs: wettability, IFT, microfluidic, and core flooding assessments. *J Mol Liq.* 2021;323:115037.
45. Ishak SA, Ghazali R, Abd Maurad Z, Zolkarnain N. Ecotoxicology study of various homologues of methyl ester sulfonates (MES) derived from palm oil. *J Surfactants Deterg.* 2017;20(6):1467–1473.
46. Tripathy DB, Mishra A, Clark J, Farmer T. Synthesis, chemistry, physicochemical properties and industrial applications of amino acid surfactants: a review. *Comptes Rendus Chimie.* 2018;21(2):112–130. doi: 10.1016/j.crci.2017.11.005
47. Akinari A, Asakura K, Osanai S. No title. *J Surf Deterg.* 2004;7(3):297–303.
48. Brito RO, Marques EF, Silva SG, do Vale ML, Gomes P, Araújo MJ, Rodriguez-Borges JE, Infante MR, Garcia MT, Ribosa I. Physicochemical and toxicological properties of novel amino acid-based amphiphiles and their spontaneously formed catanionic vesicles. *Colloids Surf B Biointerfaces.* 2009;72(1):80–87.
49. Rosa M, del Carmen Morán M, da Graça Miguel M, Lindman B. The association of DNA and stable catanionic amino acid-based vesicles. *Colloids Surf A Physicochem Eng Asp.* 2007;301(1–3):361–375.
50. Hirasaki GJ, Miller CA, Puerto M. 2011. Recent Advances in Surfactant EOR. (December). doi:10.2118/115386-PA
51. Pacwa-Płociniczak M, Płaza GA, Piotrowska-Seget Z, Cameotra SS. Environmental applications of biosurfactants: recent advances. *Int J Mol Sci.* 2011;12(1):633–654. doi: 10.3390/ijms12010633
52. Roy S, Dey J. Self-organization properties and microstructures of sodium N-(11-acrylamidoundecanoyl)-L-valinate and-L-threoninate in water. *Bull Chem Soc Jpn.* 2006;79(1):59–66.
53. Rodríguez-Escontrela I, Rodríguez-Palmeiro I, Rodríguez O, Arce A, Soto A. Characterization and phase behavior of the surfactant ionic liquid tributylmethylphosphonium dodecylsulfate for enhanced oil recovery. *Fluid Phase Equilib.* 2016;417:87–95.
54. Zana R. Critical micellization concentration of surfactants in aqueous solution and free energy of micellization. *Langmuir.* 1996;12(5):1208–1211.
55. Ray GB, Ghosh S, Moulik SP. Physicochemical studies on the interfacial and bulk behaviors of sodium N-dodecanoyl sarcosinate (SDDS). *J Surfactants Deterg.* 2009; 12(2):131–143.
56. Zhang Y, You Q, Fu Y, Zhao M, Fan H, Liu Y, Dai C. Investigation on interfacial/surface properties of bio-based surfactant N-aliphatic amide-N, N-diethoxypropylsulfonate sodium as an oil displacement agent regenerated from waste cooking oil. *J Mol Liq.* 2016;223:68–74.
57. Kronberg B, Holmberg K, Lindman B. *Surface chemistry of surfactants and polymers.* [place unknown]: John Wiley & Sons; 2014.
58. Tackie-Otoo BN. 2023. EXPERIMENTAL INVESTIGATION OF THE OIL RECOVERY POTENTIALS OF ALTERNATIVE ASP FORMULATIONS FOR SANDSTONE AND CARBONATE RESERVOIRS.
59. Pinheiro L, Faustino C. Amino acid-based surfactants for biomedical applications. *Application and Characterization of Surfactants.* 2017:207–232.

60. Seo S, Mastiani M, Mosavati B, Peters DM, Mandin P, Kim M. Performance evaluation of environmentally benign nonionic biosurfactant for enhanced oil recovery. *Fuel.* 2018;234:48–55.
61. Marshak S. *Earth: portrait of a planet: 5th International Student Edition.* [place unknown]: WW Norton & Company; 2015.
62. Tackie-Otoo BN, Atta DY, Ayoub Mohammed MA, Otchere DA. Investigation into the oil recovery process using an organic alkali–amino acid-based surfactant system. *Energy & Fuels [Internet].* 2021;35:1171–11192. doi: 10.1021/acs.energyfuels.1c00551
63. Dong S, Li Y, Song Y, Zhi L. Synthesis, characterization and performance of unsaturated long-chain carboxybetaine and hydroxy sulfobetaine. *J Surfactants Deterg.* 2013;16(4):523–529.
64. Muggeridge A, Cockin A, Webb K, Frampton H, Collins I, Moulds T, Salino P. Recovery rates, enhanced oil recovery and technological limits. *Philos Trans Royal Society A: Math Phys Eng Sci.* 2014;372(2006):20120320.
65. Bennett B, Buckman JO, Bowler BFJ, Larter SR. Wettability alteration in petroleum systems: the role of polar non-hydrocarbons. *Pet Geosci.* 2004;10(3):271–277.
66. Yuan Y, Lee TR. 2013. Contact Angle and Wetting Properties. [place unknown]. doi: 10.1007/978-3-642-34243-1
67. SalehiM,JohnsonSJ,LiangJ-T.Mechanisticstudyofwettabilityalterationusingsurfactants with applications in naturally fractured reservoirs. *Langmuir.* 2008;24(24):14099–14107. doi: 10.1021/la802464u
68. Sheng J. *Modern chemical enhanced oil recovery: theory and practice.* [place unknown]: Gulf Professional Publishing; 2010.
69. Saxena N, Kumar S, Mandal A. Adsorption characteristics and kinetics of synthesized anionic surfactant and polymeric surfactant on sand surface for application in enhanced oil recovery. *Asia-Pac J Chem Eng.* 2018;13(4):e2211.
70. Goudie KJ, McCreath SJ, Parkinson JA, Davidson CM, Liggat JJ. Investigation of the influence of pH on the properties and morphology of gelatin hydrogels. *J Polym Sci.* 2023;61(19):2316–2332.
71. Ye, Z., Zhang, F., Han, L., Luo, P., Yang, J., & Chen, H. (2008). The effect of temperature on the interfacial tension between crude oil and gemini surfactant solution. *Colloids and Surfaces A: Physicochemical and Engineering Aspects*, *322*(1), 138–141. https://doi.org/https://doi.org/10.1016/j.colsurfa.2008.02.043

7 Amino Acids' Role in Geo CO_2 Capture

Daniel Ekow Yalley, Amirun Nissa Rehman and Johannes Ami

7.1 INTRODUCTION

Carbon dioxide (CO_2) geological capture, an essential strategy in mitigating climate change, is increasingly pivotal in contemporary discussions surrounding environmental sustainability.[1,2] Industries, particularly those in oil and gas production, grapple with the imperative to curtail CO_2 emissions, necessitating efficient storage solutions.[3] While conventional techniques such as geological storage in depleted reservoirs and saline aquifers exist, uncertainties regarding their efficacy and potential environmental repercussions persist.[4]

In response to these challenges, hydrate-based CO_2 storage has emerged as a promising avenue, offering significant potential for effective and safe CO_2 sequestration.[5–10] This innovative approach capitalizes on the formation of solid crystalline structures, or gas hydrates, wherein CO_2 molecules are entrapped within water cages under suitable thermodynamic conditions.[11,12] Hydrate-based storage presents notable advantages, including high storage capacity, with approximately 120–180 m^3 of CO_2 potentially storable in 1 m^3 of hydrate at standard conditions.[13,14]

However, despite its promise, hydrate-based CO_2 storage confronts various challenges, notably the slow kinetics of hydrate formation and concerns regarding stability, particularly within sediments.[6–10] Chemical additives, such as NaCl and SDS, have been explored to enhance kinetics, with amino acid-based additives showing particular promise.[15] Nevertheless, a comprehensive understanding of the influence of additives, especially amino acids, on hydrate formation kinetics and stability, particularly in the presence of brine, remains a critical research gap.[16]

Furthermore, the role of sediment properties in influencing hydrate formation and dissociation kinetics necessitates thorough investigation to optimize CO_2 storage in porous media.[17,18] While studies have examined the kinetics of CO_2 hydrate formation in various porous media, including silica sand and clay, further research is needed to elucidate the influence of sediment properties on hydrate stability and formation mechanisms.[10,17,19–21]

DOI: 10.1201/9781003502388-7

7.2 CO_2 CAPTURE IN POROUS MEDIA

Carbon dioxide storage is the current focus of most research studies to curb the huge CO_2 emissions from industries. There have been several attempts to develop ways to store the emitted CO_2 in geological locations in various forms. The reason for choosing these locations is to be able to store large amounts of CO_2. The next section describes briefly the limitations of various existing conventional CO_2 storage techniques and the challenges encountered in their implementation.

7.3 GAS HYDRATE GEO CO_2 CAPTURE TECHNIQUES

The current carbon capture and storage (CCS) solutions involve injecting CO_2 collected from large sources into deep geological formations. These technologies include enhanced oil recovery (EOR), depleted gas reservoirs, non-mineable coal seams, and promising basalts, as shown in Figure 7.1.

However, according to the state of the technology, CCS requires a lot of energy, roughly 30% of which is required for the capture process. Furthermore, a considerable amount of CO_2 is released for every kilogram of CO_2 that is held.

Numerous CO_2 emission reduction solutions rely on EOR applications in which CO_2 is substituted with oil during recovery.[23–30] However, for successful storage, critical elements such as reservoir depth, pressure, temperature, and storage capacity must be correctly maintained to prevent significant financial costs.[31] Furthermore,

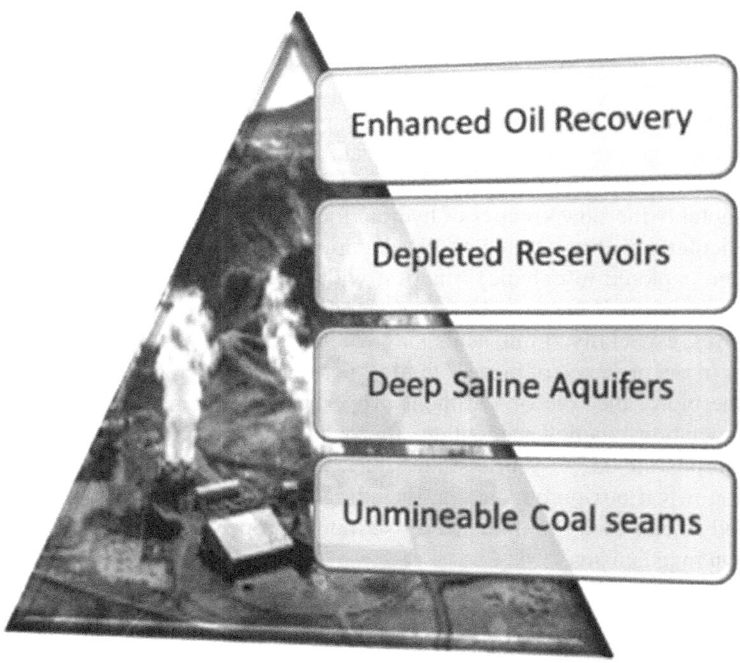

FIGURE 7.1 Geological carbon dioxide storage techniques[22].

CO_2-EOR combined with CCS technology has been shown to be thermodynamically impractical because the process consumes more energy than it produces.[32] Depleted reservoirs are another type of often-used geological sequestration technology for CO_2 storage.[33] Storage efficiency in saline aquifers is determined by a number of factors, including aquifer characteristics such as pressure, temperature, water salinity, porosity, permeability, heterogeneity, and CO_2 injection rate,[33] as well as viscous forces, capillary forces, and buoyancy caused by the density difference between injected CO_2 and brine. This difference in density causes CO_2 to migrate upward,[34] followed by gravity segregation and subsequently moving to the surface, resulting in leakage.[35]

CO_2 storage is a major issue faced around the globe where the conventional storage techniques fail to provide stable long-term storage due to leakage concerns. Thus, there arises the need to look for potential CO_2 storage methods that could serve as a substitute and provide stable storage without any leakage. In this regard, the past few years have witnessed extensive research on hydrate-based methods for CO_2 storage.

7.4 HYDRATE-BASED CO_2 CAPTURE IN POROUS MEDIA

This section addresses the kinetics and thermodynamics of CO_2 hydrate formation in porous media with reference to potential CO_2 storage applications. Details of the hydrate-based CO_2 storage investigation in several kinds of porous media are shown in Table 7.1 and Figure 7.2. The CO_2 hydrate-related porous media studies are either based on CO_2 sequestration or storage, depending on the application. Nonetheless, the characteristics and impact of the porous medium type on the kinetics of CO_2 hydrate formation (hydrate formation duration, rate, and storage) are equally relevant for both purposes. The authors typically investigate the kinetics and thermodynamics of CO_2 hydrate production in several porous media systems with diverse characteristics. The purpose of the kinetic research is to comprehend how hydrate formation in porous media behaves over time. However, the thermodynamic studies offer relevant details about the ideal circumstances for storing CO_2 as hydrates. The porous media must be under high pressure and low temperature in order to store CO_2 as hydrates. In coastal conditions, sandstones are a useful porous medium for storing CO_2 as hydrate. Known as silicates, these sandstones are typically composed of feldspar or quartz. The sandstone grains are held together by the presence of cementing materials such as clays or shales. Grain diameters in sandstones typically range from 62.5 to 2,000 μm. Although it is challenging to obtain real marine cored sandstone samples, authors often investigate porous media minerals that are commonly found in sandstones. In the literature, several authors have employed porous media for lab-scale investigations, including silica sand, silica gel, kaolin, bentonite clay, quartz sand, Toyoura sand, cellulose foam, glass beads, and metallic packing. There is currently no literature on CO_2 hydrate porous media studies using mixed porous media systems. However, for commercialization purposes, using sandstone core samples for testing might produce better results. In addition, some authors choose porous media primarily in order to achieve appropriate rock qualities that are

TABLE 7.1
Summary of Reported CO_2 Hydrate Studies in Porous Media

Study Type	Type of Porous Media	Additives	Results and Limitations	Author
THS/KHS	Na-MMT	H_2O +NaCl	Hydrates are thermodynamically inhibited by clay minerals and salts. No research on the dissociation of hydrates.	42
KHS	Silica gel	Tween-80, DTACI and SDS	SDS raises the initial hydrate formation rate and improves hydrate formation absorption. No research on the dissociation of hydrates.	36
KHS	Silica gel	DI	Higher consumption of gas was the outcome of silica gel's increased surface area. The best substance for increasing the rate of hydrate formation was SDS. Hydrate dissociation with brine has not been studied.	37
THS	Silica gel	DI	An inhibitory effect occurs when pore diameter increases because it lowers the water activity in silica gel pores. No brine.	51
KHS	Kaolin and Na-MMT	DI	CO_2 hydrates are thermodynamically inhibited by bentonite. Heat transmission regulates the development of hydrate in kaolinite. Not brine.	44
KHS	Silica sand, Zeolite	SDS	In silica sand, SDS encourages the development of hydrate more than zeolite does. No study on brine-induced hydrate dissociation.	43
KHS	Silica sand	DI	The storage capacity of CO_2 hydrate in porous medium is increased by an appropriate CO_2 injection technique. Not brine.	39
KHS	Quartz sand	DI	Hydrate of CO_2 particle size decreases lead to an increase in storage capacity. Not brine.	18
KHS	Silica sand	Seawater	The smaller the silica particle, the more gas is used. Salts reduce petrol consumption, which inhibits thermodynamics. No research on the dissociation of hydrates	17

(Continued)

TABLE 7.1
Continued

Study Type	Type of Porous Media	Additives	Results and Limitations	Author
KHS	Quartz sand	DI	The average formation rate and maximal gas storage capacity are found in porous media with pore sizes of 13.8 nm. The gas storage capacity of the porous media increases with decreasing pore size. No research on the dissociation of hydrates with brine.	18
KHS	Pumice, FHRC, and silica sand	DI	The hydrate formation kinetics of pumice were superior to those of FHRC. Lowering the bed height improved kinetics. Hydrate formation kinetics were improved by smaller particle sizes. No study on brine-induced hydrate dissociation.	20
THS	Soda glass, mix glass beads and silica gel	DI	Hydrates are thermodynamically inhibited by decreasing pore size because of the capillary effect. No brine.	52
KHS	Cellulose foam	DI	More hydrates are formed by lower water saturation than by higher saturation. No brine.	53
KHS	Glass beads	DI	Hydrates form in two stages—hydrate enclathration and continuous occupancy. Contrary to the preceding continuous stage, the gas occupancy stage lasted longer. Hydrate dissociation with brine has not been studied.	40
KHS	Toyoura sand	NaCl	In saltwater, the rate at which hydrates form initially is higher. Compared to pure water, saltwater showed a somewhat lower rate of water conversion to hydrate. No research has been done on how particle size affects the kinetics of hydrate formation and dissociation.	19
KHS	Silica sand and silica gel	DI	Compared to silica gel, silica sand has a greater water conversion rate. No research on brine.	21

(Continued)

TABLE 7.1
Continued

Study Type	Type of Porous Media	Additives	Results and Limitations	Author
KHS	Glass beads	DI	A key element in the creation of CO_2 hydrates is flow rate. High CO_2 storage necessitates an ideal flow rate. Hydrate formation is hindered by low operating pressure. No research on brine.	41
THS/KHS	Glass beads	THF SDS	With a shorter induction period, the driving force for hydrate formation increases as the concentration of THF rises. The hydrate phase equilibrium pressure is significantly reduced by THF. Brine has not been studied.	52
KHS	Glass beads	THF	Hydrate formation is enhanced by high operating pressure. No research on brine.	54
KHS	Silica gel	DI	The driving force affects how rapidly hydrates form, with 100 nm silica gel holes displaying the fastest rate of production. Hydrate dissociation with brine has not been studied.	48
THS	Silica gel	DI	The generation of CO_2 hydrate is thermodynamically affected by the silica gel's pore size. No research in brine.	49
KHS	Silica sand, silica gel, metallic packing	SDS	As the initial pressure got higher, so did the rate of hydrate formation. Hydrate dissociation with brine is yet to be studied.	58
KHS	Frozen quartz sand	DI	As the initial pressure got higher, so did the rate of hydrate formation. Hydrate dissociation with brine is yet to be studied.	55
KHS	Silica sand	C5H10	Compared to UTR, FBR exhibited more gas uptake and a higher rate of hydrate formation. No brine.	56
KHS	Na-MMT and UBS	NaCl	Na-MMT had a shorter induction time than the others. But in the Ulleung basin sediment (UBS), hydrate nucleation occurred faster. No research on the dissociation of hydrates.	57

(Continued)

TABLE 7.1
Continued

Study Type	Type of Porous Media	Additives	Results and Limitations	Author
THS	Glass beads	Seawater	The hydrate equilibrium temperature and gas uptake go up with an increase in MCP concentration. The particle size effect has yet to be studied.	58
THS/KHS	Silica gel	THF, EGME, SDS, and TBAB	THF and SDS increase the driving force for gas uptake. Hydrate dissociation with brine has not been studied.	59
THS	Silica gel	DI	The equilibrium curve for N_2 hydrates shifts to lower pressures as the concentration of CO_2 increases. No brine.	60
THS	Glass beads	DI	At the same temperature, gas mixtures with a higher CO_2 mole fraction exhibit lower equilibrium pressure. No brine.	61
THS	Glass beads	DI	The equilibrium curve for N_2 hydrates changes to lower pressures as the concentration of CO_2 increases. No brine.	62
THS/KHS	Silica gel	DI	The rate of hydrate formation was accelerated by silica gel. No brine.	63
THS	Glass beads	SDS	The equilibrium pressure in the hydrate phase was lowered using THF. At 1000 mg/l, SDS is the highest hydrate saturation of any additive. No brine	52
THS	Metal-organic framework	DI	Only the meso and macro holes formed gas hydrates, which exhibited a minor inhibition in contrast to the bulk phase. No brine.	64
THS	Silica gel	DI	When compared to bulk water, silica gel accelerates the rate of hydrate formation and water-to-hydrate conversion. No brine.	48
THS	Crystalline swelled Na-MMT	DI	On the hydrate phase boundary curve, gas hydrates generated in the clay exhibit both promotion and inhibitory effects. No brine	65

(Continued)

TABLE 7.1
Continued

Study Type	Type of Porous Media	Additives	Results and Limitations	Author
THS	Quartz sand and silica gel	NaCl	In contrast to bulk hydrate, the dissociation curve of mixed CO_2 + CH_4 hydrate in porous media moved to the left. At a constant temperature, equilibrium pressure dropped as the gas phase's CO_2 concentration rose.	66
THS/KHS	Soda glass and silica gel	THF and SDS	At 3 mol%, THF improves gas consumption. Gas usage rises as driving force increases. No brine.	61
THS	Silica glass	DI	Pore diameter has an impact on hydrate dissociation temp. A lower decomposition temperature was the outcome of increased capillary pressure. No brine.	67
THS	Na-MMT, kaolinite, and illite	Glycine, glucose, and urea	While kaolinite and illite had no effect, sodium montmorillonite had an impact on CO_2 hydrate phase equilibrium. The equilibrium of the CO_2 hydrate phase is inhibited by organic materials.	57

Abbreviations: THS = thermodynamic hydrate study; KHS = kinetic hydrate study; THF = tetrahydrofuran; SDS = sodium dodecylsulfate; TBAB = tetrabutylammonium bromide; FHRC = fire-hardened red clay; Na-MMT = sodium montmorillonite/bentonite.

comparable to those of sandstones. This is mostly accomplished by employing porous media with varying grain or particle sizes. The grain size range employed to examine hydrate formation in porous media is 0.0007–1,000 μm. At the same time, the porosity ranged from 0.15 to 0.559 and was investigated at various water saturations.

Pure CO_2,[18,38–43] binary mixes of CO_2 + CH_4[45–49] and CO_2+N_2,[50] and additional guest molecules such as C_3H_8, C_2H_6, THF, TBAB, and cyclopentane have all been examined as hydrate formers. Depending on the guest molecule mixes, the experimental pressures and temperatures ranged from 1.13 to 10.50 MPa and 269.65 to 289.95 K. The inclusion of various guest molecules was done to see how they affected CO_2 hydrate formation in porous media by changing the phase boundary curve or improving the kinetics. In addition, other scientists investigated the effects of hydration promoters like SDS, amino acids, Tween-80, DTACl, urea, glycine, and glucose. These experiments are generally carried out in brine solution to simulate the seawater influence in marine sediments. The additives are investigated at concentrations less

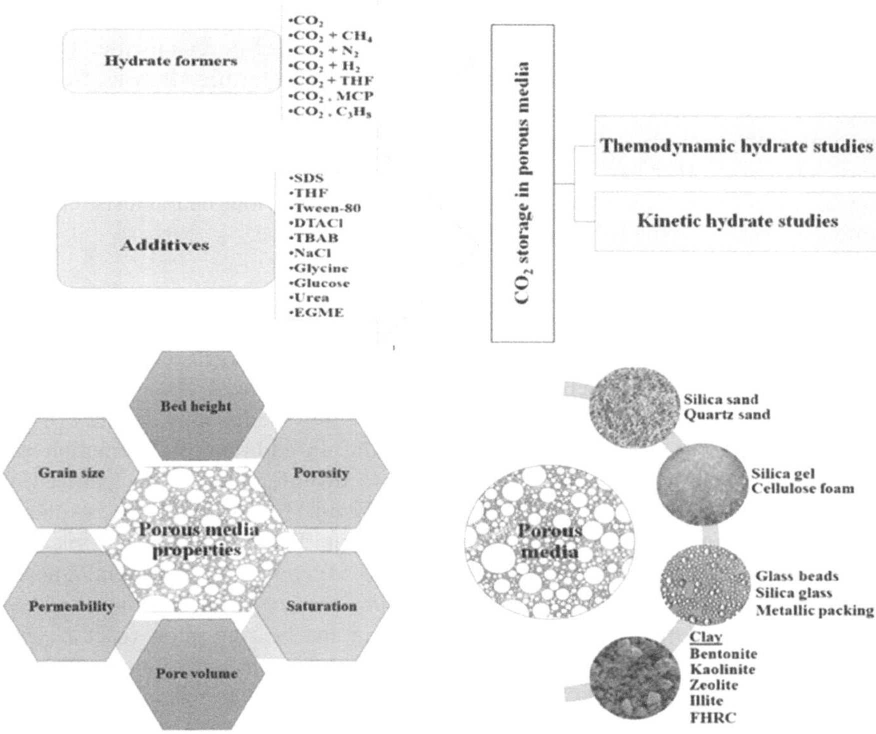

FIGURE 7.2 Overview of CO_2 hydrate research in various porous media systems.

than 20 wt%, depending on their qualities. By far the bulk of research undertaken use the standard hydrate autoclave cell to investigate hydrate formation in porous material. Other researchers added MRI and NMR analyzers to their experimental sets to perform advanced analyses.

7.5 HYDRATE PHASE BEHAVIOR IN POROUS MEDIA

Determining the hydrate phase equilibrium conditions is essential in creating practical recommendations for CO_2 storage in porous media. The stability of the hydrates and their ability to store CO_2 can be assessed by having an adequate knowledge of the CO_2 hydrate phase boundary conditions. Determining the storage capacity is important for sequestering CO_2 in subsea reservoirs in order to predict the prospective amount of CO_2 that could potentially form. Insights into the potential CO_2 storage margin are provided by the practical thermodynamic conditions appropriate for storing CO_2 under seabed circumstances. However, in order to assess the effectiveness of CO_2 separation in fuel gas and flue systems for carbon capture applications, the phase behavior of CO_2 in porous media is frequently examined.

Understanding the behavior of the CO_2 hydrate phase in fuel gas and flue gas offers important details about CO_2 separation and selectivity in porous media. The hydrate phase behavior curve is strongly influenced by the kind of porous media. The size and distribution of the particles in any porous medium determine how it affects the hydrate phase boundary. Because it controls the system's surface area and capillary effect, the particle size of porous medium is important. Particle size also has a quasi-relation to other parameters like porosity and permeability, which may affect how hydrates form. This is vital to understand how the various characteristics of the porous media affect the behavior of the CO_2 hydrate phase.

7.6 HYDRATE KINETICS IN POROUS MEDIA

The kinetics of hydrate formation in porous media has been suggested for the understanding of the formation behavior of storing CO_2 as hydrates. As a consequence, such kinetic indicators can offer important insights into the limiting quantity of CO_2 that can be sequestered within any porous media. In addition, the formation rate of the hydrate, the amount of CO_2 converted to hydrate, and the nucleation time of hydrate in porous media could be well analyzed. To elucidate this, exploring the impact of each kinetic parameter on CO_2 hydrate formation in porous media is required. However, hydrate kinetic data or results are apparatus dependent, ignoring nature of experimental conditions like temperature, pressure, concentration, and type of porous media. Other factors including cooling and heating rates, reactor types, reactor shape, reactor wall roughness, and the gas injection flow rate can all affect the kinetic measurement. Furthermore, the kinetics of hydrate formation in porous media can be influenced by a number of different circumstances. The hydrate formation mechanism in porous media can be greatly influenced by the type of additives used, their concentration, particle size, water to gas saturation (bed height), porosity, and the type of porous medium. Consequently, these factors must be carefully considered when discussing the effect of hydrate formation in porous media. The last factor contributing to uncertainty in these measurements is that hydrates are formed stochastically. We are thus confronted with a formidable challenge: how to estimate the kinetics of hydrate formation.

7.7 HYDRATE BASED IN AMINO ACIDS SYSTEMS

Amino acids are biologically organic compounds that are the building blocks of protein.

Naturally, 20 types of amino acids occur in living organisms. The general structure of amino acid consists of an amine ($-NH_2$) and carboxylic acid ($-COOH$) as functional groups with a unique side chain (ranging from a polar alkyl chain (hydrophobic) to a positively or negatively charged moiety (hydrophilic)). These unique side chains strongly influence the physical and chemical properties of the amino acids.[68–70] Figure 7.3 shows the general structure of amino acids.

Amino acids can be classified based on the side chain group (R-group) due to the difference in their physicochemical properties. In this study, the R-group of the amino acids studied vary from aliphatic (isoleucine), sulfur-containing (methionine)

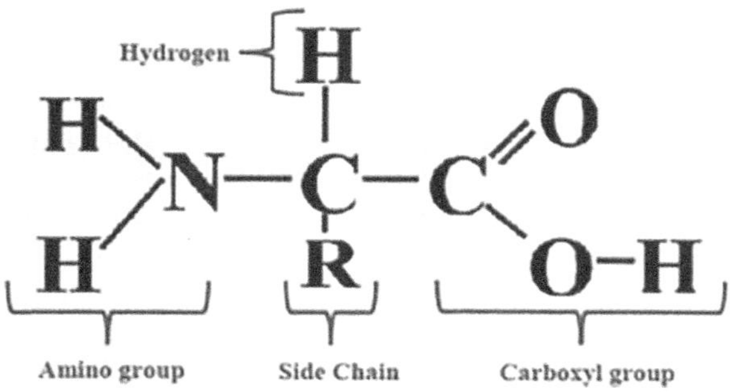

FIGURE 7.3 General amino acid structure[71].

TABLE 7.2
Structure and Physicochemical Properties of Studied Amino Acids

Amino Acid	Structure	Side Chain Polarity	Hydro phobicity	Side Chain
L-Methionine	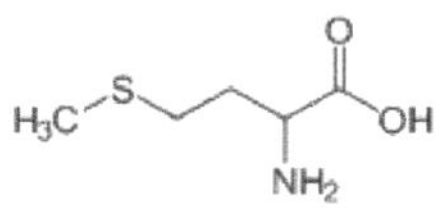	Polar, noncharged	1.9	CH3-S-(CH2)2-
L-Isoleucine	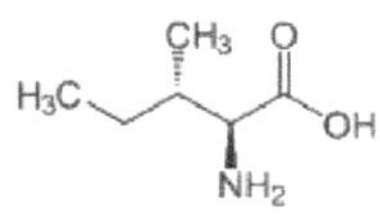	Nonpolar	4.5	–CH(CH3) C2H5
L-Threonine	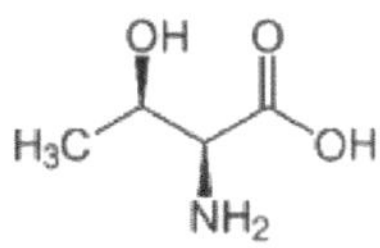	Polar, noncharged	–0.7	CH3-CH(OH)

to alcoholic (threonine) group. Table 7.2 presents the structure and the physicochemical properties of the studied amino acids in this work.

7.8 AMINO ACID AND WATER INTERACTION

The effect of amino acids on hydrogen-bonded water network reorientation dynamics is critical in understanding their role in the gas hydrate promotion mechanism. The different behavior of amino acids in aqueous solutions arises from their unique side chain group since amino acids have the same basic structure. The affinity of amino acids for water molecules has been studied based on different physicochemical properties. These studies have resulted in developing an amino acid side chain hydrophobic scale. There are several such scales[72] since authors study different properties and propose amino acid hydrophobicity based on their results. Generally, authors use the amino acid hydrophobic scale proposed by Kyte and Doolittle.[73]

However, the reason for this selection scale is not known. In addition, there are several contributions on the amino acids effect of water dynamics. The water reorientation dynamics mechanism is very important during the hydrate formation process as water molecules form hydrogen-bonded cage-like networks that trap the guest molecules (gas). Hence, the presence of hydrophilic groups like carboxyl and amino groups in amino acid structure helps to form strong interactions with the water molecules via hydrogen bonding.

7.9 AMINO ACID AS GAS HYDRATE PROMOTERS

Amino acids as kinetic promoters for CO_2 hydrate were first reported by Cai et al.[74] using L-methionine. They studied hydrate formation kinetics at different concentrations of L-methionine in aqueous solutions and reported the optimal performance of L-methionine at 0.2 wt% showing enhanced CO_2 uptake. Kinetic promotion of CO_2 hydrates was also reported by Prasad and Kiran[75] using L-valine, L-phenylalanine, L-cysteine, L-methionine, and L-threonine at 0.5 wt% using two different reactor configurations. Their results suggested L-methionine, L-cysteine, and L-valine as best showing faster and higher gas uptake under static conditions. The individual and synergistic effect of amino acids L-methionine and L-phenylalanine on CO_2, CH_4, and (CO_2+CH4) mixed gas systems showed enhanced kinetics with rapid hydrate conversions at 0.5 wt%.[76] Another study by Lui et al.[77] showed kinetic promotion of CO_2 hydrate using L-isoleucine at 0.2 wt% with high CO_2 uptake. The promotion kinetics of CO_2 hydrates was reported using different concentrations (0, 100, 300, and 1,000 ppm) of L-tryptophan in a stirred tank reactor. They concluded 300 ppm of L-tryptophan as the optimum concentration exhibiting less induction time and high CO_2 uptake. Amino acid-related studies concerning geological CO_2 sequestration in porous media are less investigated in the literature. Pandey et al.[78] studied CO_2 hydrate formation kinetics in porous media in the presence of promoters L-valine, L-methionine, and L-histidine at 3,000 ppm. When compared with the commercial promoter SDS, the amino acids were found to enhance the hydrate formation kinetics. Their findings revealed the promotion effect of L-histidine over L-methionine, L-valine, and SDS based on the induction time measurements. Another study by Pandey et al.[79] reports amino acid kinetics in porous media using CO_2+N_2 gas mixtures for CO_2 capture, separation, and storage applications. They used liquid CO_2 in a rocking cell apparatus to investigate the hydrate kinetics using quartz sand as porous media, and amino acids— L-methionine, L-valine, $_{L\text{-}}$histidine, L-arginine, and SDS. The results showed higher gas uptake and higher moles of CO_2 recovered in presence of the amino acids, suggesting the amino acids are a potential replacement for conventional promoters like SDS for CO_2 capture and storage applications.

However, all these studies are conducted in pure water and do not represent the actual subsurface condition that contains brine. There is limited literature on the effect of amino acids on CO_2 hydrate formation and dissociation kinetics in the presence of brine. The kinetic data gathered are considered differently since the kinetics of gas hydrate formation is stochastic in nature and depends on several factors such as apparatus type, reactor height, wall thickness, driving force, and reactor wall roughness.[80] Table 7.3 shows the obtained kinetic data from various studied systems

TABLE 7.3
Experimental Conditions and Gas Uptake Measurements of Studied Amino Acids as Kinetic Hydrate Promoters

No.	Amino Acid	Conc. (wt%)	Gas	*P* (MPa)	*T* (K)	Gas Uptake	Reference
1	L-Norvaline	0.2	CO_2	3.3	273.2	361 mg/g	74
2	L-Norleucine	0.2	CO_2	3.3	273.2	362 mg/g	74
3	2- Aminoheptanoic acid	0.2	CO_2	3.3	273.2	55 mg/g	74
4	*n*-Hexanoic acid	0.2	CO_2	3.3	273.2	56 mg/g	74
5	*n*-Hexylamine	0.2	CO_2	3.3	273.2	135 mg/g	74
6	L-Methionine	0.2	CO_2	3.3	273.2	350 mg/g	74
7	L-Methionine	0.02	CO_2	3.3	273.2	350 mg/g	74
8	L-Methionine	0.05	CO_2	3.3	273.2	355 mg/g	74
9	L-Methionine	0.1	CO_2	3.3	273.2	356 mg/g	74
10	L-Methionine	0.2	CO_2	3.3	273.2	346 mg/g	74
11	L-Methionine	0.5	CO_2	3.3	273.2	321 mg/g	74
12	L-Valine	1.0	CO_2	3.3	298	61.93 mmol/mol H2O	75
13	L-Cysteine	0.5	CO_2	3.3	298	52.10 mmol/mol H2O	75
14	L-Methionine	0.5	CO_2	3.3	298	70.74 mmol/mol H2O	75
15	L-Methionine	0.5	CH4				76
16	L-Phenylalanine	0.5	CH4				76
17	L-Phenylalanine	0.5	CO_2+CH4				76
18	L-Methionine	0.5	CO_2+CH4				76
19	L-Isoleucine	0.1	CO_2	3.3	273.2	335 mg/g	77
20	L-Isoleucine	0.2	CO_2	3.3	273.2	331 mg/g	77
21	L-Isoleucine	0.5	CO_2	3.3	273.2	330 mg/g	77
22	L-Isoleucine	1.0	CO_2	3.3	273.2	290 mg/g	77
23	L-Tryptophan	0	CO_2	3.4	273.65	0.020	81
24	L-Tryptophan	0.01	CO_2	3.4	273.65	0.019	81
25	L-Tryptophan	0.03	CO_2	3.4	273.65	0.014	81
26	L-Tryptophan	0.1	CO_2	3.4	273.65	0.019	81
27	L-Tryptophan	0	CO_2	3.6	275.65	0.019	81
28	L-Tryptophan	0.01	CO_2	3.6	275.65	0.021	81
29	L-Tryptophan	0.03	CO_2	3.6	275.65	0.019	81
30	L-Tryptophan	0.1	CO_2	3.6	275.65	0.022	81
31	L-Tryptophan	0	CO_2	3.8	277.65	0.021	81
32	L-Tryptophan	0.01	CO_2	3.8	277.65	0.021	81
33	L-Tryptophan	0.03	CO_2	3.8	277.65	0.021	81
34	L-Tryptophan	0.1	CO_2	3.8	277.65	0.019	81

in the literature. Generally, the kinetic promotion indicator evaluated includes gas uptake (Table 7.3) and induction time. High gas uptake with less induction time is usually preferred to achieve maximum efficiency of the CO_2 hydrate storage process under subsurface conditions.

7.10 EFFECT OF AMINO ACID SOLUBILITY AND CONCENTRATION

The solubility of amino acids is an important phenomenon that determines the hydrate forming potential of amino acids. Several factors affect the solubility of amino acids such as the shape, size, hydrogen-bonding capacity, hydrophobicity, hydrophilicity, and zwitterionic character.[82] The interaction of amino acids with water molecules varies based on the concentration as well as the properties of the unique side chain attached. Thus, the hydrate formation kinetics is influenced by the hydrogen bonding activity of the amino acid with the water molecules. However, before hydrate formation, it is important to ensure the solubility of amino acids. The type of solvent plays a critical role in governing the solubility of amino acids. The presence of an electrolyte affects the solubility of amino acids.[83] For practical applications of CO_2 hydrate storage under subsurface conditions, considering the effect of brine on the solubility of amino acids is highly important. In an aqueous solution, amino acids exist as zwitterions with a negatively charged carboxyl group and a positively charged amino group. These charged groups create a strong electrostatic field around the amino acid molecules that interact with the ions in the electrolyte solution resulting in important ion-amino acid interactions. These interactions could be long-range interactions causing salting-out effect or in the form of ion-pair complexes causing salting-in effect based on the structure of the amino acid and at times the electrolyte concentration. For example, Roy et al.[82] reported a salting-out effect for L-tyrosine and L-leucine while a salting-in effect was observed for L-cysteine. Similar variations in solubility of different amino acids resulting from amino acid and electrolyte interaction were reported by Khoshkbarchi and Vera.[83] Therefore, a clear understanding of these interactions is necessary to gain insight into the effect of amino acids on CO_2 hydrate formation behavior in the brine system.

7.11 AMINO ACIDS KINETIC PROMOTION MECHANISM

Studies in literature report the kinetic promotion mechanism of amino acids based on the reaction of amino acids with CO_2 molecules (known as zwitterionic mechanism) and the structure-property relationship.[71,74,77] Amino acids possess different functional groups and hydrophobicity which affects the interaction of the amino acids with the solvent molecules. These interactions depend on the type of solvent and the concentration of amino acids. In case of salt solutions such as brine, the concentration of the salt also influences the amino acid and solvent interactions. The chain length in hydrophobic amino acids is an important factor that determines the promotion or inhibition behavior of amino acids during hydrate formation. According to Cai et al.,[74] there exist an optimum alkyl chain length in hydrophobic amino acids that promotes CO_2 hydrate formation. They confirmed this in their study using five different amino acids where very long and short alkyl chain lengths showed hydrate inhibition. Thus, the promotion mechanism of L-methionine arises from the optimum alkyl chain length and the combined effect of the hydrophilic carboxyl and amino groups present. Furthermore, Liu et al.[77] proposed a similar kinetic promotion mechanism based on the chain length and the number of methyl groups attached to the amino acid molecule. The kinetic promotion mechanism of amino acids has been reported in literature for methane hydrates based on varying concentrations of amino acid where the hydrates formed are flexible and porous in nature.[84] The porous nature of the hydrates causes more water to migrate from the bulk solution to the gas-liquid interface through the capillary suction mechanism, leading to kinetic promotion effect. They also concluded that the concentration of the amino acid solution plays a key role during the hydrate formation mechanism. It has been observed that higher amino acid concentrations (>1 wt%) inhibit natural gas hydrate formation while lower concentrations (≤1 wt%) promote hydrate formation kinetics.

However, all the kinetic promotion mechanisms reported in the literature are conducted in pure water which is different from this study that considers amino acid solutions in brine (3.3 wt% NaCl solution). The mechanism of kinetic promotion for CO_2 hydrate formation depends on the interaction of amino acids with brine. The solubility of amino acids is affected by the concentration of electrolyte and the nature of both the amino acid and the cation of the electrolyte. For example, L-threonine is highly soluble in an electrolyte solution of NaCl (brine) due to the presence of the –OH group in the side chain that increases the polarity of L-threonine increasing its tendency to dissolve in an ionic solution. This results in a salting-in effect whereby the solubility of the amino acid (L-threonine) is increased. However, in the case of L-isoleucine salting-out effect was observed resulting from the presence of two methyl ($-CH_3$) groups that create more pronounced hydrophobic interactions.[83]

7.12 KINETIC MODELING OF CO_2 HYDRATES IN POROUS MEDIA

There is no gas-hydrate-based kinetic modeling in porous media in the presence of amino acids with brine reported in the literature to the best of our knowledge. However, kinetic modeling of CO_2 hydrates in porous media has been reported in the literature with regard to CO_2 sequestration applications. Palodkar and Jana[85] formulated a mathematical model to interpret the growth and decomposition mechanism of methane and CO_2 hydrates in presence of silica sand porous media and seawater, mimicking the marine environment. A theoretical formulation has been proposed by Palodkar and Jana[86] to understand the characteristics of methane and CO_2 swapping phenomena that occur in permafrost and under deep seabed considering chemical potential as the driving force. Palodkar and Jana[87] proposed a kinetic model based on the chemical potential difference for CO_2 hydrate formation in porous media. The formation and growth kinetics of CO_2 hydrates have been modeled in sand sediments and reservoir conditions considering various factors such as surface area, the morphology of hydrates, pore size, shape of internal pores as well as the effect of salt ions that influence the hydrate kinetics in porous media (Table 2.5). Simulation studies have been performed using different software such as Material Studio[88] and TOUGH+HYDRATE simulators[89] as shown in Table 7.3.

However, no study has been reported that uses classical nucleation theory (CNT) to predict the hydrate formation kinetics with amino acids in porous media. CNT has been commonly used to predict the rate of nucleation and the height of the free energy barrier. Since hydrate formation is a crystallization process, CNT can be employed in predicting hydrate formation kinetics in porous media systems. The CNT was originally derived for the condensation of vapor into liquid in the 1920s, and it has been later applied to the crystallization of supersaturated solutions by analogy. Due to its analytical simplicity, it has been widely applied not just for crystallization but also for clathrate hydrates. This work adopts the CNT model to predict carbon dioxide formation nucleation time in porous media with amino acid-based promoters.

7.13 CONCLUSION

The exploration into the utilization of amino acids as promoters for CO_2 hydrate formation presents a compelling avenue for advancing CCS technologies. Through an extensive review of literature, we have delved into various aspects of amino acid kinetics, solubility, and their potential mechanisms in enhancing CO_2 uptake within hydrate formations.

The studies showcased the promising performance of amino acids such as L-methionine, L-valine, and others in promoting CO_2 hydrate formation, with notable improvements in kinetics and gas uptake observed under different experimental conditions. These findings underscore the potential of amino acids as efficient alternatives to conventional promoters like SDS, particularly in enhancing CO_2 storage efficiency under subsurface conditions.

Moreover, the discussions on amino acid solubility, concentration effects, and their interaction with brine highlight the complexity of hydrate formation dynamics in real-world scenarios. Understanding these factors is crucial for accurately modeling CO_2 hydrate formation kinetics, especially in porous media environments, where brine presence significantly influences amino acid behavior and hydrate formation processes. While the literature provides valuable insights into amino acid kinetics and their impact on CO_2 hydrate formation, there exists a notable gap in kinetic modeling, particularly in porous media with brine conditions. Bridging this gap through the incorporation of CNT holds promise for developing predictive models that can guide the design of more efficient CCS strategies.

In essence, the research presented today not only underscores the potential of amino acids as kinetic promoters for CO_2 hydrate formation but also emphasizes the importance of further investigations into their behavior in realistic subsurface environments. Such endeavors are crucial for advancing our understanding of CO_2 storage mechanisms and unlocking the full potential of hydrate-based CCS technologies in mitigating greenhouse gas emissions.

REFERENCES

1. Mukherjee A, Okolie JA, Abdelrasoul A, Niu C, Dalai AK. Review of post-combustion carbon dioxide capture technologies using activated carbon. *J Environ Sci (China)*. 2019;83:46–63. doi:10.1016/j.jes.2019.03.014

2. Graus W, Blomen E, Worrell E. Global energy efficiency improvement in the long term: a demand- and supply-side perspective. *Energy Effic.* 2011;4(3):435–463. doi:10.1007/s12053-010-9097-z
3. Bavoh CB, Partoon B, Lal B, Gonfa G, Foo Khor S, Sharif AM. Inhibition effect of amino acids on carbon dioxide hydrate. *Chem Eng Sci.* 2017;171:331–339. doi:10.1016/j.ces.2017.05.046
4. Mukherjee A, Okolie JA, Abdelrasoul A, Niu C, Dalai AK. Review of post-combustion carbon dioxide capture technologies using activated carbon. *J Environ Sci.* 2019;83:46–63. doi:10.1016/J.JES.2019.03.014
5. Dashti H, Zhehao Yew L, Lou X. Recent advances in gas hydrate-based CO2 capture. *J Nat Gas Sci Eng.* 2015;23:195–207. doi:10.1016/j.jngse.2015.01.033
6. Zheng J, Chong ZR, Qureshi MF, Linga P. Carbon Dioxide Sequestration via Gas Hydrates: a potential pathway toward Decarbonization. *Energy and Fuels.* 2020;34(9):10529–10546. doi:10.1021/acs.energyfuels.0c02309
7. Kvamme B, Graue A, Buanes T, Kuznetsova T, Ersland G. Storage of CO2 in natural gas hydrate reservoirs and the effect of hydrate as an extra sealing in cold aquifers. *Int J Greenh Gas Con.* 2007;1(2):236–246. doi:10.1016/S1750-5836(06)00002-8
8. Gauteplass J, Almenningen S, Ersland G, Barth T, Yang J, Chapoy A. Multiscale investigation of CO2 hydrate self-sealing potential for carbon geo-sequestration. *Chem Eng J.* 2020;381(August 2019):122646. doi:10.1016/j.cej.2019.122646
9. Tohidi B, Yang J, Salehabadi M, Anderson R, Chapoy A. CO2 hydrates could provide secondary safety factor in subsurface sequestration of CO2. *Environ Sci Technol.* 2010;44(4):1509–1514. doi:10.1021/es902450j
10. Rehman AN, Pendyala R, Lal B. Effect of brine on the kinetics of Carbon dioxide hydrate formation and dissociation in porous media. *Mater Today Proc.* 2020;47(xxxx):1366–1370. doi:10.1016/j.matpr.2021.04.024
11. Kassim Z, Saad Khan M, Lal B. Thermodynamic modelling on methane hydrate equilibrium condition in the presence of electrolyte inhibitor. *Mater Today Proc.* 2019;19:1395–1402. doi:10.1016/j.matpr.2019.11.158
12. Khan MS, Bavoh CB, Partoon B, Lal B, Bustam MA, Shariff AM. Thermodynamic effect of ammonium based ionic liquids on CO2 hydrates phase boundary. *J Mol Liq.* 2017;238:533–539. doi:10.1016/j.molliq.2017.05.045
13. Koh CA, Sloan ED, Sum AK, Wu DT. Fundamentals and applications of gas hydrates. *Annu Rev Chem Biomol Eng.* 2011;2:237–257. doi:10.1146/annurev-chembioeng-061010-114152
14. Wang X, Zhang F, Lipiński W. Research progress and challenges in hydrate-based carbon dioxide capture applications. *Appl Energy.* 2020;269(May):114928. doi:10.1016/j.apenergy.2020.114928
15. Sa JH, Kwak GH, Lee BR, Park DH, Han K, Lee KH. Hydrophobic amino acids as a new class of kinetic inhibitors for gas hydrate formation. *Sci Rep.* 2013;3:1–7. doi:10.1038/srep02428
16. Qanbari F, Pooladi-Darvish M, Tabatabaie SH, Gerami S. CO2 disposal as hydrate in ocean sediments. *J Nat Gas Sci Eng.* 2012;8:139–149. doi:10.1016/j.jngse.2011.10.006
17. Mekala P, Busch M, Mech D, Patel RS, Sangwai JS. Effect of silica sand size on the formation kinetics of CO2hydrate in porous media in the presence of pure water and seawater relevant for CO2sequestration. *J Pet Sci Eng.* 2014;122:1–9. doi:10.1016/j.petrol.2014.08.017
18. Zhang X, Li J, Wu Q, Wang C, Nan J. Experimental study on the effect of pore size on carbon dioxide hydrate formation and storage in porous media. *J Nat Gas Sci Eng.* 2015;25:297–302. doi:10.1016/j.jngse.2015.05.014
19. Yang SHB, Babu P, Chua SFS, Linga P. Carbon dioxide hydrate kinetics in porous media with and without salts. *Appl Energy.* 2016;162:1131–1140. doi:10.1016/j.apenergy.2014.11.052

20. Bhattacharjee G, Kumar A, Sakpal T, Kumar R. Carbon dioxide sequestration: influence of porous media on hydrate formation kinetics. *ACS Sustain Chem Eng.* 2015;3(6):1205–1214. doi:10.1021/acssuschemeng.5b00171
21. Babu P, Kumar R, Linga P. Pre-combustion capture of carbon dioxide in a fixed bed reactor using the clathrate hydrate process. *Energy.* 2013;50(1):364–373. doi:10.1016/j.energy.2012.10.046
22. Rehman AN. 2023. Kinetics And Stability Studies Of Co2 Hydrate In Quartz Sand And Amino Acids For CO2 Storage [Internet]. [place unknown]: UNIVERSITI TEKNOLOGI PETRONAS; [accessed 2025 Mar 28]. https://utpedia.utp.edu.my/id/eprint/27687
23. Jia W, McPherson BJ, Pan F, Xiao T, Bromhal G. Probabilistic analysis of CO2 storage mechanisms in a CO2-EOR field using polynomial chaos expansion. *Int J Greenh Gas Con.* 2016;51:218–229. doi:10.1016/J.IJGGC.2016.05.024
24. Damiani D, Litynski JT, McIlvried HG, Vikara DM, Srivastava RD. The US Department of Energy's R&D program to reduce greenhouse gas emissions through beneficial uses of carbon dioxide. *Greenh Gases: Sci Technol.* 2012;2(1):9–16. doi:10.1002/ghg.35
25. Bachu S, Shaw J. Evaluation of the CO2 Sequestration Capacity in Alberta's Oil and Gas Reservoirs at Depletion and the Effect of Underlying Aquifers. *J Can Pet Technol.* 2003;42(9):51–61. doi:10.2118/03-09-02
26. Qi R, LaForce TC, Blunt MJ. 2008. Design of carbon dioxide storage in oilfields. *Proceedings - SPE Annual Technical Conference and Exhibition.* 3(September):1730–1741. doi:10.2118/115663-ms
27. Trivedi JJ, Babadagli T. Oil recovery and sequestration potential of naturally fractured reservoirs during CO2 injection. *Energy and Fuels.* 2009;23(8):4025–4036. doi:10.1021/ef900361n
28. Gorecki CD. 2012. Integrating CO2 EOR and CO2 storage in the Bell Creek oil field. *Carbon Management Technology Conference* 2012.:507–520. doi:10.7122/151476-ms
29. Saini D, Jimenez I. 2014. Evaluation of CO2-EOR and storage potential in mature oil reservoirs. *Society of Petroleum Engineers Western North America and Rocky Mountain Joint Conference and Exhibition* 2014. 2(April):673–684. doi:10.2118/169555-ms
30. Wang H, Liao X, Dou X, Shang B, Ye H, Zhao D, Liao C, Chen X. Potential evaluation of CO2 sequestration and enhanced oil recovery of low permeability reservoir in the Junggar Basin, China. *Energy and Fuels.* 2014;28(5):3281–3291. doi:10.1021/ef4025068
31. Bachu S. Identification of oil reservoirs suitable for CO2-EOR and CO2 storage (CCUS) using reserves databases, with application to Alberta, Canada. *Int J Greenh Gas Con.* 2016;44:152–165. doi:10.1016/J.IJGGC.2015.11.013
32. Farajzadeh R, Eftekhari AA, Dafnomilis G, Lake LW, Bruining J. On the sustainability of CO2 storage through CO2 – Enhanced oil recovery. *Appl Energy.* 2020;261:114467. doi:10.1016/J.APENERGY.2019.114467
33. Celia MA, Bachu S, Nordbotten JM, Bandilla KW. Status of CO2 storage in deep saline aquifers with emphasis on modeling approaches and practical simulations. *Water Resour Res.* 2015;51(9):6846–6892. doi:10.1002/2015WR017609
34. Ajayi T, Gomes JS, Bera A. A review of CO2 storage in geological formations emphasizing modeling, monitoring and capacity estimation approaches. *Pet Sci.* 2019;16(5):1028–1063. doi:10.1007/s12182-019-0340-8
35. Hesse MA, Tchelepi HA, Orr FM. 2006. Scaling analysis of the migration of CO2 in saline aquifers. *Proceedings - SPE Annual Technical Conference and Exhibition.* 5:3001–3012. doi:10.2118/102796-ms
36. Kang SP, Lee JW. Kinetic behaviors of CO2 hydrates in porous media and effect of kinetic promoter on the formation kinetics. *Chem Eng Sci.* 2010;65(5):1840–1845. doi:10.1016/j.ces.2009.11.027

37. Kumar A, Sakpal T, Linga P, Kumar R. Influence of contact medium and surfactants on carbon dioxide clathrate hydrate kinetics. *Fuel.* 2013;105:664–671. doi:10.1016/j.fuel.2012.10.031
38. Kumar A, Sakpal T, Linga P, Kumar R. Enhanced carbon dioxide hydrate formation kinetics in a fixed bed reactor filled with metallic packing. *Chem Eng Sci.* 2015;122:78–85. doi:10.1016/j.ces.2014.09.019
39. Sun D, Englezos P. Storage of CO2 in a partially water saturated porous medium at gas hydrate formation conditions. *Int J Greenh Gas Con.* 2014;25:1–8. doi:10.1016/j.ijggc.2014.03.008
40. Zheng J nan, Yang L, Ma S, Zhao Y, Yang M. Quantitative analysis of CO2 hydrate formation in porous media by proton NMR. *AIChE Journal.* 2020;66(2):1–9. doi:10.1002/aic.16820
41. Yang M, Song Y, Jiang L, Zhu N, Liu Y, Zhao Y, Dou B, Li Q. CO2 hydrate formation and dissociation in cooled porous media: a potential technology for CO2 capture and storage. *Environ Sci Technol.* 2013;47(17):9739–9746. doi:10.1021/es401536w
42. Lee JW, Chun MK, Lee KM, Kim YJ, Lee H. Phase equilibria and kinetic behavior of CO2 hydrate in electrolyte and porous media solutions: application to ocean sequestration of CO2. *Korean J Chem Eng.* 2002;19(4):673–678. doi:10.1007/BF02699316
43. Arora A, Kumar A, Bhattacharjee G, Kumar P, Balomajumder C. Effect of different fixed bed media on the performance of sodium dodecyl sulfate for hydrate based CO2 capture. *Mater Des.* 2015;90:1186–1191. doi:10.1016/j.matdes.2015.06.049
44. Zhang Y, Li X, Chen Z, Cai J, Xu C, Li G. Formation behaviors of CO2 Hydrate in Kaoline and Bentonite Clays with Partially Water Saturated. *Energy Procedia.* 2017;143:547–552. doi:10.1016/j.egypro.2017.12.724
45. Palodkar AV, Jana AK. Clathrate hydrate dynamics with synthetic- and bio-surfactant in porous media: model formulation and validation. *Chem Eng Sci.* 2020;213:115386. doi:10.1016/j.ces.2019.115386
46. Stevens JC, Baldwin BA, Graue A, Ersland G, Husebo J, Howard JJ. Measurements of hydrate formation in sandstone. *Petrophysics.* 2008;49(1):67–73.
47. Song Y, Zhou H, Ma S, Liu W, Yang M. CO2 sequestration in depleted methane hydrate deposits with excess water. *Int J Energy Res.* 2018;42(7):2536–2547. doi:10.1002/er.4042
48. Kang SP, Seo Y, Jang W. Kinetics of methane and carbon dioxide hydrate formation in silica gel pores. *Energy and Fuels.* 2009;23(7):3711–3715. doi:10.1021/ef900256f
49. Seo Y, Lee H, Uchida T. Methane and carbon dioxide hydrate phase behavior in small porous silica gels: three-phase equilibrium determination and thermodynamic modeling. *Langmuir.* 2002;18(24):9164–9170. doi:10.1021/la0257844
50. Nakashima T, Sato T, Inui M. Numerical modeling of hydrate formation in sand sediment simulating sub-seabed CO2 storage in the form of gas hydrate. *Energy Procedia.* 2013;37:5986–5993. doi:10.1016/j.egypro.2013.06.526
51. Kang SP, Lee JW, Ryu HJ. Phase behavior of methane and carbon dioxide hydrates in meso- and macro-sized porous media. *Fluid Phase Equilib.* 2008;274(1–2):68–72. doi:10.1016/j.fluid.2008.09.003
52. Yang M, Song Y, Liu W, Zhao J, Ruan X, Jiang L, Li Q. Effects of additive mixtures (THF/SDS) on carbon dioxide hydrate formation and dissociation in porous media. *Chem Eng Sci.* 2013;90:69–76. doi:10.1016/j.ces.2012.11.026
53. Nambiar A, Babu P, Linga P. CO2 capture using the clathrate hydrate process employing cellulose foam as a porous media. *Can J Chem.* 2015;93(8):808–814. doi:10.1139/cjc-2014-0547
54. Yang M, Zhou H, Wang P, Li N, Song Y. Hydrate-based CO2 capture from flue gas in constant pressure process with the presence of THF. *Energy Procedia.* 2017;142:3939–3943. doi:10.1016/j.egypro.2017.12.300

55. Zhang X, Li J, Wu Q, Wang Y, Wang J, Li Y. Effect of initial pressure on the formation of Carbon Dioxide Hydrate in Frozen Quartz Sand. *Energy and Fuels*. 2019;33(11):11346–11352. doi:10.1021/acs.energyfuels.9b01693
56. Zheng J, Zhang BY, Wu Q, Linga P. Kinetic Evaluation of Cyclopentane as a Promoter for CO2 Capture via a Clathrate Process Employing Different Contact Modes. *ACS Sustain Chem Eng*. 2018;6(9):11913–11921. doi:10.1021/acssuschemeng.8b02187
57. Park T, Kyung D, Lee W. Effect of organic matter on CO2 hydrate phase equilibrium in phyllosilicate suspensions. *Environ Sci Technol*. 2014;48(12):6597–6603. doi:10.1021/es405099z
58. Zheng JN, Yang M. Phase Equilibrium Data of CO2-MCP Hydrates and CO2 Gas Uptake Comparisons with CO2-CP Hydrates and CO2-C3H8 Hydrates. *J Chem Eng Data*. 2019;64(1):372–379. doi:10.1021/acs.jced.8b00893
59. Abu Hassan MH, Sher F, Zarren G, Suleiman N, Tahir AA, Snape CE. Kinetic and thermodynamic evaluation of effective combined promoters for CO2 hydrate formation. *J Nat Gas Sci Eng*. 2020;78(September 2019):103313. doi:10.1016/j.jngse.2020.103313
60. Seo YT, Moudrakovski IL, Ripmeester JA, Lee JW, Lee H. Efficient recovery of CO2 from flue gas by clathrate hydrate formation in porous silica gels. *Environ Sci Technol*. 2005;39(7):2315–2319. doi:10.1021/es049269z
61. Song Y, Wang X, Yang M, Jiang L, Liu Y, Dou B, Zhao J, Wang S. Study of selected factors affecting hydrate-based carbon dioxide separation from simulated fuel gas in porous media. *Energy and Fuels*. 2013;27(6):3341–3348. doi:10.1021/ef400257a
62. Yang M, Song Y, Liu Y, Jiang L, Zhao Y. 2016. Thermodynamic Characters of N 2 / CO2 Hydrates in Marine Sediment. :880653.
63. Kang SP, Lee J, Seo Y. Pre-combustion capture of CO2 by gas hydrate formation in silica gel pore structure. *Chem Eng J*. 2013;218:126–132. doi:10.1016/j.cej.2012.11.131
64. Kim D, Ahn YH, Lee H. Phase Equilibria of CO2 and CH4 Hydrates in Intergranular Meso/Macro Pores of MIL-53 Metal Organic Framework. *J Chem Eng Data*. 2015;60(7):2178–2185. doi:10.1021/acs.jced.5b00322
65. Kim D, Ahn YH, Kim SJ, Lee JY, Lee J, Seo YJ, Lee H. Gas Hydrate in Crystalline-Swelled Clay: the effect of Pore Dimension on Hydrate Formation and Phase Equilibria. *J Phys Chem. C*. 2015;119(38):22148–22153. doi:10.1021/acs.jpcc.5b03229
66. Mu L, Cui Q. Experimental study on the dissociation Equilibrium of (CH4 + CO2) Hydrates in the (Quartz Sands + NaCl Solution) *System. J Chem Eng Data*. 2019;64(12):6041–6048. doi:10.1021/acs.jced.9b00859
67. Anderson R, Llamedo M, Tohidi B, Burgass RW. Experimental measurement of methane and carbon dioxide clathrate hydrate equilibria in mesoporous silica. *J Phys Chem. B*. 2003;107(15):3507–3514. doi:10.1021/jp0263370
68. Vaitheeswaran S, Thirumalai D. Interactions between amino acid side chains in cylindrical hydrophobic nanopores with applications to peptide stability. *Proc Natl Acad Sci U S A*. 2008;105(46):17636–17641. doi:10.1073/pnas.0803990105
69. Madeira PP, Bessa A, Álvares-Ribeiro L, Raquel Aires-Barros M, Rodrigues AE, Uversky VN, Zaslavsky BY. Amino acid/water interactions study: a new amino acid scale. *J Biomol Struct Dyn*. 2014;32(6):959–968. doi:10.1080/07391102.2013.800994
70. Vyas N, Ojha AK. Interaction of alanine with small water clusters; Ala-(H2O)n (n = 1, 2 and 3): A density functional study. *J Mol Struct: THEOCHEM*. 2010;940(1–3):95–102. doi:10.1016/j.theochem.2009.10.015
71. Bavoh CB, Lal B, Osei H, Sabil KM, Mukhtar H. A review on the role of amino acids in gas hydrate inhibition, CO2 capture and sequestration, and natural gas storage. *J Nat Gas Sci Eng*. 2019;64(November 2018):52–71. doi:10.1016/j.jngse.2019.01.020

72. Wilce MCJ, Aguilar MI, Hearn MTW. Physicochemical basis of Amino Acid Hydrophobicity Scales: evaluation of Four New Scales of Amino Acid Hydrophobicity Coefficients Derived from RP-HPLC of Peptides. *Anal Chem.* 1995;67(7):1210–1219. doi:10.1021/ac00103a012
73. Kyte J, Doolittle RF. A simple method for displaying the hydropathic character of a protein. *J Mol Biol.* 1982;157(1):105–132. doi:10.1016/0022-2836(82)90515-0
74. Cai Y, Chen Y, Li Q, Li L, Huang H, Wang S, Wang W. CO2 Hydrate formation Promoted by a Natural Amino Acid l-Methionine for Possible Application to CO2 Capture and Storage. *Energy Technol.* 2017;5(8):1195–1199. doi:10.1002/ente.201600731
75. Prasad PSR, Kiran BS. Are the amino acids thermodynamic inhibitors or kinetic promoters for carbon dioxide hydrates? *J Nat Gas Sci Eng.* 2018;52:461–466. doi:10.1016/j.jngse.2018.02.001
76. Prasad PSR, Kiran BS. Synergistic effects of amino acids in clathrates hydrates: gas capture and storage applications. *Chem Eng J Adva.* 2020;3(July):100022. doi:10.1016/j.ceja.2020.100022
77. Liu Z, Zeng Y, Wang W. CO2 Hydrate formation Promoted by a Bio-friendly Amino Acid L-Isoleucine. *IOP Conf Ser Earth Environ Sci.* 2020;474(5):052054. doi:10.1088/1755-1315/474/5/052054
78. Pandey JS, Daas YJ, Von Solms N. *CO2 Storage in Sediments by Hydrate Formation and Self-Preservation in the Presence of Promoters.* AAPG Europe Regional Conference; 2020.
79. Pandey JS, Daas YJ, Karcz AP, Solms N von. Enhanced hydrate-based geological CO2 capture and sequestration as a mitigation strategy to address climate change. *Energies (Basel).* 2020;13(21):1–28. doi:10.3390/en13215661
80. Sloan D. Clathrate Hydrates of Natural Gases. *Bull Geol Soc.* 1991;41:80–85.
81. Khandelwal H, Qureshi MF, Zheng J, Venkataraman P, Barckholtz TA, Mhadeshwar AB, Linga P. Effect of l -Tryptophan in Promoting the Kinetics of Carbon Dioxide Hydrate Formation. *Energy and Fuels.* 2021;35(1):649–658. doi:10.1021/acs.energyfuels.0c03709
82. Roy S, Guin PS, Mahali K, Dolui BK. Amino acid solubility under the influence of NaCl at 298.15 K. *J Mol Liq.* 2016;218:316–318. doi:10.1016/j.molliq.2016.02.054
83. Khoshkbarchi MK, Vera JH. Effect of NaCl and KCl on the solubility of amino acids in aqueous solutions at 298.2 K: measurements and modelling. *Ind. Eng. Chem. Res.* 1997;36(6):2445–2451. https://doi.org/10.1021/ie9606395
84. Veluswamy HP, Lee PY, Premasinghe K, Linga P. Effect of Biofriendly Amino Acids on the Kinetics of Methane Hydrate Formation and Dissociation. *Ind Eng Chem Res.* 2017;56(21):6145–6154. doi:10.1021/acs.iecr.7b00427
85. Palodkar AV, Jana AK. Growth and Decomposition Mechanism of Clathrate Hydrates in the Presence of Porous Media and Seawater: Experimental Validation. *Energy and Fuels.* 2019;33(2):1433–1443. doi:10.1021/acs.energyfuels.8b03245
86. Palodkar AV, Jana AK. Fundamental of swapping phenomena in naturally occurring gas hydrates. *Sci Rep.* 2018;8(1):1–10. doi:10.1038/s41598-018-34926-2
87. Palodkar AV, Mandal S, Jana AK. Modeling Growth Kinetics of Gas Hydrate in Porous Media: Experimental Validation. *Energy and Fuels.* 2016;30(9):7656–7665. doi:10.1021/acs.energyfuels.6b01397
88. Ding T, Liu Y. Simulations and analysis of hydrate formation during CO2 injection into cold saline aquifers. *Energy Procedia.* 2014;63:3030–3040. doi:10.1016/j.egypro.2014.11.326
89. Yu T, Sato T, Nakashima T, Inui M, Oyama H. An integrated model for CO2 hydrate formation in sand sediments for sub-seabed CO2 storage. *Int J Greenh Gas Con.* 2016;52:250–269. doi:10.1016/j.ijggc.2016.07.009

8 Amino Acids' Role in H_2 Geo-Storage

Collins Mensah, Solomon Adjei Marfo and Frank Bonah-Okyere

8.1 INTRODUCTION

Hydrogen is currently evaluated as a promising clean fuel in the hydrogen economy. However, a significant challenge in realizing a full-scale hydrogen economy lies in hydrogen storage, given its high compressibility and volatility. One potential solution is hydrogen geo-storage, where compressed hydrogen is injected into geological formations, allowing for on-demand withdrawal.[1] The hydrogen economy has garnered substantial attention from both industry and academia, as it offers a sustainable energy distribution system. Hydrogen serves not as a primary energy source but as an energy carrier produced from a variety of resources through diverse energy conversion processes.[2] The proposed H_2 economy envisions widespread production and use of hydrogen as a primary energy carrier.[3] For this economy to flourish, safe and cost-effective hydrogen storage is imperative.[3] The underground gas storage concept emerged to meet consumer demand during high seasonal peaks, utilizing geological formations to accommodate large hydrogen quantities and to stabilize fluctuations in demand. Furthermore, it assists with the storage and regeneration of power from intermittent renewable sources such as wind and solar.[2] Hydrogen can be stored above ground as a refrigerated liquid, in metal hydrides, as a high-pressure gas, or underground in salt caverns, aquifers, and other formations.[4] The most desirable method for hydrogen storage is as a solid hydrate, due to its relatively low pressure and temperature requirements for storage and retrieval. However, the slow kinetics of hydrate formation necessitates the use of kinetic hydrate promoters (KHPs) to enhance the process. Recently, hydrophobic amino acids like L-methionine, L-valine, and L-leucine have gained attention as environmentally friendly and economical KHPs for gas hydrate systems.[5]

8.2 H_2 GEO STORAGE TECHNIQUES

Various underground storage methods include salt caverns, mined caverns, depleted hydrocarbon fields, and aquifers (porous media). Among these, the salt cavern technique is favored for underground hydrogen storage (UHS) due to its advantages, such as high withdrawal capacity, low construction costs, and increased safety.[6] Methods for storing compressed air and hydrogen gas in underground salt caverns show promise for large-scale energy storage applications.[2]

 DOI: 10.1201/9781003502388-8

8.2.1 Salt Cavern Storage

Salt caverns are developed within naturally occurring thick salt domes located deep underground, with global availability except in certain Pacific Rim areas.[7] They are created through a process known as solution mining or leaching, which involves the controlled injection of water to dissolve salt, subsequently removing the brine via a single well. The same well is then used for gas injection and withdrawal.[6] For instance, a salt cavern about 1,000 m deep and with a geometrical storage volume of 700,000 m^3 can hold around 6 million kg of hydrogen at maximum operating pressure.[2] Successful solution mining and cavern design hinge on appropriate placement and engineering within salt deposits. The basic leaching techniques consist of direct and indirect leaching. In direct leaching, freshwater is introduced into the cavern via an inner string, while for indirect leaching, it flows through the outer string. A combination of these methods is often implemented, utilizing a protective blanket medium lighter than water to prevent salt dissolution in the roof area.[8] Upon completion of the leaching process and confirming that the cavern meets design specifications, leaching strings are replaced with gas withdrawal equipment, and the cavern transitions into a pressurized gas storage facility.[8] Hydrogen is then withdrawn through pressure reduction and reintroduced into the gas pipeline using a separator and dehydration unit.[2]

8.2.2 Storage in Mined Rock Caverns

Mined rock caverns consist of underground chambers excavated within rock masses such as limestone, chalk, or granite. The storage principle is hydraulic confinement, with caverns constructed at depths permitting sufficient hydrostatic pressure to counteract the vapor pressure of the stored hydrogen.[9]

8.2.3 Porous Media Storage (Depleted Fields or Aquifers)

This involves using voids in sedimentary rocks, characterized by high porosity and permeability, to store gas. The porosity indicates the amount of available space, while permeability reflects the ease of gas flow through the rock formation. Suitable storage sites may include deep aquifers with high salinity or depleted oil and gas reservoirs, which are shielded by poorly permeable cap rocks to prevent leakage.[10]

8.2.4 Hydrate-Based Hydrogen Storage

Using the solid-hydrate form for hydrogen storage is advantageous due to its requirements for moderate temperature and pressure, mitigating the energy demands of liquid hydrogen. Yet, slow kinetics remain a challenge for developing this storage method on a larger scale.[5] Hydrates form under conditions of high pressure and low temperature and have previously been considered problematic in oil and gas production; however, interest has shifted toward their potential utility for future fuel storage.[11] Gas hydrate promoters enhance formation rates, and current research looks toward biodegradable alternatives such as amino acids, which provide eco-friendly kinetic action without foam formation—an issue with traditional surfactant-based KHPs.[11] Studies suggest that the kinetic promotion effects depend on factors like amino acid side chain length and hydropathy.[12,13]

8.2.5 Liquid Organic Hydrogen Carriers

Hydrogen storage using liquid organic hydrogen carrier (LOHC) involves reversible hydrogenation and dehydrogenation of unsaturated molecules.[14] This process facilitates transport, allowing the hydrogen to be released at the destination. Though LOHC holds higher gravimetric storage density compared to metal hydrides and compressed gas, the associated high temperatures and pressures for reactions present disadvantages (Tables 8.1 and 8.2).

8.3 H_2 STORAGE IN AMINO ACID SYSTEMS

Hydrogen storage technology is critical for the advancement of sustainable energy solutions, and two primary methodologies are employed for this purpose: storage as hydrates and the utilization of LOHCs. The hydrate technique is predominantly utilized for UHS, given its unique properties, while the LOHC method is favored for its ability to facilitate the safe transportation of hydrogen, as these carriers are generally nonflammable. In the LOHC storage process, hydrogenation is employed to chemically bind hydrogen within the carrier, enabling its subsequent release through dehydrogenation when required. To enhance the kinetics of hydrate-based storage systems, amino acids have been introduced as additives. A significant challenge associated with the LOHC approach is the substantial thermal energy required for the dehydration process. However, certain amino alcohols, notably 2-amino-ethanol,[15] have been identified as effective in reducing this thermal demand, highlighting the need for further investigation in this area due to the limited existing literature. Recent studies indicate the synergistic potential of coupling thermodynamic promoters with kinetic promoters to enhance the kinetics of hydrogen hydrate formation. For instance, Zhang et al.[5] successfully coupled L-valine with tetrahydrofuran (THF)

TABLE 8.1
Compilation of the Standard Molar Enthalpies of Vaporization $\Delta^g_l H^\circ_m$ of Substituted Amino-Alcohols

Compound	Temperature Range	Standard Molar Enthalpies of Vaporization, $\Delta^g_l H^\circ_m$	Reference
2-amino-ethanol	341–453	57.9 ± 1.5	17
2-(methyl-amino)-ethanol	298.2–308.2	57.5 ± 1.7	18
2-(dimethyl-amino)-ethanol	336–409	46.4 ± 0.8	16
2-(diethyl-amino)-ethanol	315–436	59.1 ± 1.2	16
1-amino-2-propanol	276.4–316.4	59.0 ± 0.4	16
3-amino-1-propanol	333–461	62.9 ± 1.3	16

TABLE 8.2
Amino Acids Used as KHP for Hydrogen and Their Properties

Amino Acid	Storage Method	Side Chain	Side Chain Polarity	Hydropathy Index	Remarks	Reference
L-Leucine	Hydrate based	$(CH3)_2$-CH–CH_2–	Hydrophobic	3.8	Effective KHP for CO_2/H_2/TBAF semi clathrate hydrate formation	19
L-Tryptophan	Hydrate based	Ph–NH–CH=C–CH_2–	Hydrophobic	–0.9	Shows excellent kinetic promotion activity for CO_2/H_2/TBAF semi-clathrate hydrate formation	20
L-Valine	Hydrate based	$(CH3)_2$–CH–	Hydrophobic	4.2	Shows excellent kinetic promotion activity for H_2/TBAF semi-clathrate hydrate formation	5
2-Amino-ethanol	LOHC				Reduces the enthalpy of dehydration of LOHC for H_2 storage	15

at a concentration of 5.56 mol% to promote the kinetics of H_2 hydrate formation under conditions of 12.0 MPa and 274.2 K, varying L-Val concentrations from 0.01 to 1.0 wt%. Their research also examined the influence of cooling rates (spanning from 0.02 to 0.3 K/min) on the kinetics of the H_2-THF hydrate, identifying an optimal cooling rate. A proposed mechanism for the formation of H_2-THF hydrates across various cooling rates was articulated to elucidate the observed increase in hydrogen uptake. The findings from Zhang et al.[5] revealed superior gravimetric hydrogen storage capacity compared to alternative H_2-THF hydrate systems, thereby confirming the technical feasibility of integrating thermodynamic and kinetic promoters to accelerate hydrate kinetics while improving hydrogen storage capacity. This research underlines the vast potential for large-scale applications of sustainable hydrate-based hydrogen storage technologies. A critical conclusion drawn from these studies is that H_2 hydrates formed at 5.56 mol% THF can operate under milder thermodynamic conditions. However, the inherent slow kinetics of H_2 hydrate formation—characterized

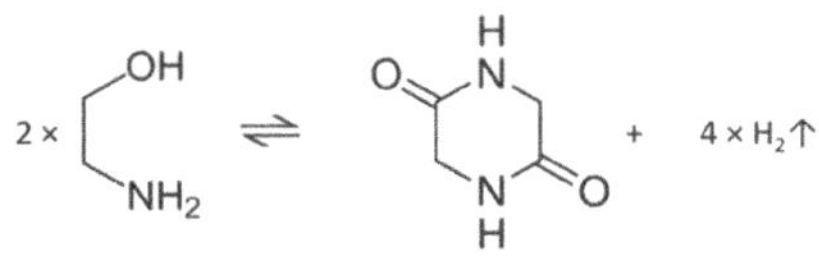

FIGURE 8.1 Reversible dehydrogenation of 2-amino-ethanol to produce piperazine-2,5-dione and H_2.

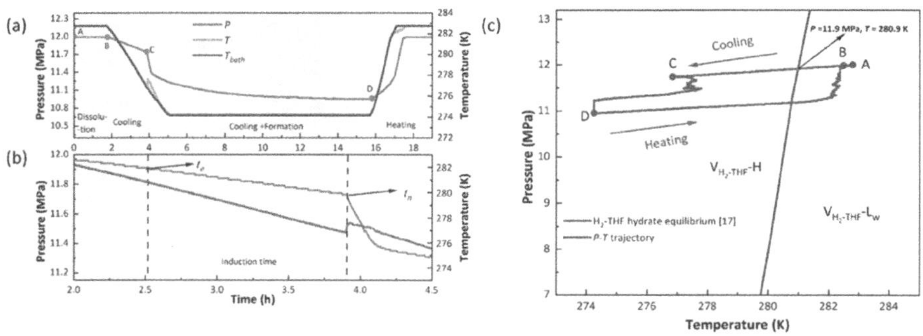

FIGURE 8.2 Pressure and temperature evolution.

by prolonged induction times and limited hydrogen storage capacities—remains a significant barrier to wider implementation. Consequently, the introduction of kinetic promoters into the H_2-THF system is posited as an effective strategy to enhance the efficacy of hydrate-based hydrogen storage. Moreover, amino alcohols have emerged as promising candidates for LOHCs in sustainable hydrogen storage applications. Verevkin et al.[16] conducted thermodynamic calculations regarding the enthalpy of the reversible dehydrogenation process of 2-amino-ethanol, revealing that it possesses a notably low enthalpy, thus marking it as a suitable and effective LOHC for the sustainable storage of hydrogen (Figure 8.1).

8.3.1 Hydrate Phase Behavior Properties of Amino Acids in H_2 System

P-T evolution and H_2 uptake during H_2-THF hydrate formation in the presence of THF 5.56 mol%

According to a study by Zhang et al.,[5] Figure 8.2 illustrates how pressure (P) and temperature (T) change during the formation and breakdown of H_2-THF hydrates. This process starts with conditions of P_0 = 12.0 MPa and T_0 = 282.6 K, with 5.56 mol% THF present. The formation happens in three steps: (1) dissolving H_2 (from point A to B), (2) cooling and forming the H_2-THF hydrate (from point B to C), and (3) growing the H_2-THF hydrate (from point C to D). After H_2 completely dissolved at T = 282.8 K, the pressure decreased slightly from 12.0 to 11.96 MPa. H_2 remained supersaturated in the THF solution until the hydrate nucleation happened, indicated by a sudden increase in temperature and a pressure drop as H_2 entered the hydrate structures. According to Cai et al.,[21] binary H_2-THF hydrates formed early in the process. The whole formation took about 14 h, with pressure dropping quickly at first and then slowing down as it approached the final point (from C to D). During the dissociation phase, the temperature was raised to 282.8 K. The P-T trajectory relating to the H_2-THF hydrate equilibrium curve is shown in Figure (b) and (c). The P-T state entered the equilibrium zone at P = 11.91 MPa and T = 280.9 K. After completing the hydrate formation, the pressure fell to 8.95 MPa, well above the equilibrium pressure, which was below 0.3 MPa.

8.3.2 Hydrate Kinetics Properties of Amino Acids in H_2 System

Zheng et al.[19] examined how leucine and tryptophan affect the speed of hydrate formation in a mixed CO_2/H_2/TBAF system. They found that using these additives at concentrations of 3,000 ppm significantly reduced the time it took to form hydrates compared to using TBAF alone. However, the rapid formation time limited how much gas could dissolve into the solution. Bhattacharjee and Linga[11] also studied how these amino acids influenced the structure of hydrate formations. They noticed different patterns when comparing formations with and without additives. In the absence of additives, many small hydrate particles were created and grew larger, which made the mixture thick and could block stirring. Without stirring, larger particles settled down visibly. In contrast, when tryptophan or leucine was used and stirring stopped right after nucleation, fewer hydrates formed, resulting in irregularly shaped chunks of hydrate and dendritic structures growing on the surfaces.

8.3.3 Mechanism of Hydrate Formation

Gas hydrate formation has two main stages: nucleation and growth. Nucleation can be challenging, especially when starting from pure water, but once hydrates are formed, the process becomes easier due to a "memory effect." It begins with tiny crystal embryos that grow. Nucleation is random and energy-driven, where only certain-sized nuclei can overcome energy barriers and grow. The transition from nucleation to growth can be hard to pinpoint. Nucleation starts with a rapid drop in pressure and continues until the hydrate growth is complete. During dissociation, the temperature is gradually increased until it synchronizes with cooling curves.

8.4 CONCLUSION

This review highlights the application of amino acids in UHS, emphasizing the potential of hydrogen as a clean energy carrier. It examines various geological formations and their suitability for different hydrogen storage scenarios. The primary techniques considered for geo-hydrogen storage include hydrate storage and the use of LOHC. However, there is limited research on these methods, resulting in a gap in the comprehensive literature. Storing hydrogen in solid hydrates presents kinetic challenges. These challenges can potentially be mitigated by incorporating specific amino acids known as KHPs, such as L-valine, L-tryptophan, and L-leucine. On the other hand, the LOHC approach faces a significant challenge due to the high energy requirement for dehydrogenation, which is on par with the energy needed for hydrogenation.

A possible solution to this energy-intensive process is the use of certain amino alcohols, like 2-amino-ethanol, which has shown promise in reducing the enthalpy required for dehydration.

REFERENCES

1. Iglauer S, Ali M, Keshavarz A. Hydrogen wettability of sandstone reservoirs: implications for hydrogen geo-storage. *Geophys Res Lett.* 2021;48(3): e2020GL090814. doi:10.1029/2020GL090814
2. Ozarslan A. Large-scale hydrogen energy storage in salt caverns. *Int J Hydrogen Energy.* 2012;37(19):14265–14277. doi:10.1016/j.ijhydene.2012.07.111
3. Aslannezhad M, Ali M, Kalantariasl A, Sayyafzadeh M, You Z, Iglauer S, Keshavarz A. A review of hydrogen/rock/brine interaction: implications for Hydrogen geo-storage. *Prog Energy Combust Sci.* 2023;95:101066. doi:10.1016/j.pecs.2022.101066
4. Davison J, Arienti S, Cotone P, Mancuso L. Co-production of hydrogen and electricity with CO2 capture. *In: Energy Procedia.* 2009;1[place unknown]:p. 4063–4070. doi:10.1016/j.egypro.2009.02.213
5. Zhang J, Li Y, Yin Z, Linga P, He T, Zheng X. Coupling amino acid L-Val with THF for superior hydrogen hydrate kinetics: implication for hydrate-based hydrogen storage. *Chem Eng J.* 2023;467:143459. doi:10.1016/J.CEJ.2023.143459
6. Tackie-Otoo BN, Haq MB. A comprehensive review on geo-storage of H2 in salt caverns: prospect and research advances. *Fuel.* 2024;356:129609. doi:10.1016/j.fuel.2023.129609
7. Shi JQ, Durucan S. CO2 Storage in caverns and mines. Oil & gas science and technology-revue d'IFP energies nouvelles. *Oil Gas Sci Technol IFP.* 2005;60(3):569–571. https://doi.org/10.2516/ogst:2005039
8. Kunstman A, Poborska-Młynarska K, Urbańczyk K. *Solution mining in salt deposits: outline of recent development trends.* [place unknown]: AGH. Uczelniane Wydawnictwa Naukowo-Dydaktyczne; 2007.
9. Nanda A. Crude Oil Storage in Large Underground Rock Caverns. [place unknown].
10. Tarkowski R, Uliasz-Misiak B, Tarkowski P. Storage of hydrogen, natural gas, and carbon dioxide – Geological and legal conditions. *Int J Hydrogen Energy.* 2021;46(38):20010–20022. doi:10.1016/J.IJHYDENE.2021.03.131
11. Bhattacharjee G, Linga P. Amino acids as kinetic promoters for gas hydrate applications: a mini review. *Energy and Fuels.* 2021;35(9):7553–7571. doi:10.1021/acs.energyfuels.1c00502
12. Madeira PP, Bessa A, Álvares-Ribeiro L, Raquel Aires-Barros M, Rodrigues AE, Uversky VN, Zaslavsky BY. Amino acid/water interactions study: a new amino acid scale. *J Biomol Struct Dyn.* 2014;32(6):959–68. doi:10.1080/07391102.2013.800994
13. Vaitheeswaran S, Thirumalai D. Interactions between amino acid side chains in cylindrical hydrophobic nanopores with applications to peptide stability. [place unknown]; 2008.
14. Niermann M, Beckendorff A, Kaltschmitt M, Bonhoff K. Liquid Organic Hydrogen Carrier (LOHC) – Assessment based on chemical and economic properties. *Int J Hydrogen Energy.* 2019;44(13):6631–6654. doi:10.1016/J.IJHYDENE.2019.01.199
15. Verevkin SP, Konnova ME, Zherikova KV, Pimerzin AA. Sustainable hydrogen storage: Thermochemistry of amino-alcohols as seminal liquid organic hydrogen carriers. *J Chem Thermodyn.* 2021;163:106591. doi:10.1016/j.jct.2021.106591
16. Verevkin SP, Andreeva IV, Konnova ME, Portnova SV, Zherikova KV, Pimerzin AA. Paving the way to the sustainable hydrogen storage: thermochemistry of amino-alcohols as precursors for liquid organic hydrogen carriers. *J Chem Thermodyn.* 2021;163:106610. doi:10.1016/j.jct.2021.106610
17. Wilson L. New Aliphatic Amines. 1935.

18. Touhara H, Okazaki S, Okino F, Tanaka H, Ikari K, Nakanishi K. Thermodynamic properties of aqueous mixtures of hydrophilic compounds 2. Aminoethanol and its methyl derivatives. *J Chem Thermodyn.* 1982;14(2):145–156. doi:10.1016/0021-9614(82)90026-X
19. Zheng J, Bhatnagar K, Khurana M, Zhang P, Zhang BY, Linga P. Semiclathrate based CO2 capture from fuel gas mixture at ambient temperature: effect of concentrations of tetra-n-butylammonium fluoride (TBAF) and kinetic additives. *Appl Energy.* 2018;217:377–389. doi:10.1016/j.apenergy.2018.02.133
20. Zhang Y, Bhattacharjee G, Linga P. Kinetic Study of Mixed Hydrogen/1,3-Dioxolane Hydrates in the Presence of Amino Acids. [place unknown]: 2021.
21. Cai J, Tao YQ, von Solms N, Xu CG, Chen ZY, Li X Sen. Experimental studies on hydrogen hydrate with tetrahydrofuran by differential scanning calorimeter and in-situ Raman. *Appl Energy.* 2019;243:1–9. doi:10.1016/j.apenergy.2019.03.179

9 Case Studies on the Use of Amino Acids in the Geo-Energy Industry

Stephen Akuoko and Ato Fanyin-Martin

9.1 INTRODUCTION

As concerns about climate change escalate, the hunt for sustainable energy sources intensifies. Geo-energy, which taps into the Earth's natural processes for energy, holds promise for a cleaner future. From geothermal power to oil recovery and carbon storage, geo-energy offers renewable options. In this quest, amino acids, basic components of life, emerge as potential game-changers. These molecules, with their unique properties and ability to break down naturally, could help solve challenges in geo-energy.[1] This paper aims to shed light on this topic by analyzing case studies showing how amino acids are used in different geo-energy applications. By studying real-world examples, we hope to uncover both the benefits and challenges of using amino acids in these contexts.

9.2 CASE STUDY 1: THE USE OF AMINO ACIDS AS HYDRATION SUPPRESSANT IN CONTROLLING REACTIVE SHALES

The pursuit of an inhibitive water-based technology that stops clay particle hydration is an ongoing project. The majority of existing techniques rely on a unique, large ionic charge strategy. Salts, like sodium or potassium chloride, are the most often utilized substances. An opposite ionic charge is used by cationic and anionic polymer systems to neutralize a clay particle's reactive sites. These high-molecular-weight polymers enclose the clay particle and preserve its physical integrity in addition to counteracting the impacts of reactive substances. This minimizes their dispersion and makes it easier to remove them from the wellbore. Even then, not all of the chemical issues relating to the inhibition of clay particles are resolved by these methods.[2]

Unfortunately, many, if not all, of the old systems have environmental shortcomings. Excessive concentrations of salts have a negative impact on the chemical and biological ecosystems. The polymer systems may pose a risk to the environment either because of their specific chemical compositions or because of their carrier phase.

 DOI: 10.1201/9781003502388-9

A recent investigation looked at components that are processed and found in nature that are connected to the physical and biological metabolism of humans. Because of their essential role in the formation of both plant and animal tissue, proteins and their basic constituents, amino acids, presented a direct challenge to this line of inquiry. One of the most intricate chemical compounds known to science, proteins are essential to every living thing's ability to function. The elements carbon, oxygen, hydrogen, nitrogen, and, in certain cases, phosphorus and sulfur are found in complex molecules called proteins. Proteins have large molecular weights, ranging from 10,000 to 20,000,000. Proteins are structurally composed of lengthy chains called peptide bonds that are connected by amide connections. Amino acids are produced when proteins are hydrolyzed.[2]

Protein hydrolysis produces amino acids, which are compounds with two groups: an amine (NH_2) and a carboxyl (COOH). While many amino acids are easily soluble in water, the majority are comparatively insoluble in organic solvents. Amino acids have characteristics more fitting of an ionic than of a covalent substance. Normally, the functional groups of amino and carboxylic acids in separate molecules combine to make a salt by further heating to remove water, which would then form an amide. The common amide formation does not take place when dealing with a single molecule that has both an amine group and a carboxylic group on nearby carbons. A single molecule has enough molecular strain due to the close spacing between its neighboring carbon atoms to stop a reaction from occurring inside the molecule. Between the acid carbon and the amine group bonded to the alpha carbon, the hydrogen of the hydroxyl group of the acid group might move. The pH affects the migration. It is usual to refer to this phenomenon as an inner salt. The ionic characteristics of amino acids are explained by this. Based on the criteria of the properties of amino acids discussed above, this case study delves into exploring the potential use of amino acids as hydration suppressants (HS) in drilling operations.

The Central Wyoming Wind River Basin served as the site of the HS's initial field test. Shale strata containing smectite present difficulties for wells in this region. Bit and bottom hole assembly balling is a common consequence of drilling with freshwater mud systems. It becomes extremely distributed ("mud-making") when drill cuttings are used; therefore, considerable water dilutions are needed to manage solids and stable mud. Methylene blue test values rise quickly, resulting in high viscosities and challenging solids removal. Although operators have opted to employ water-based fluids (WBF) treated with KCl polymer in an effort to lessen the shale issues, the wells continue to present challenges. Although oil-based fluids were formerly a possibility, their usage is declining because of environmental concerns. To drill the problematic Waltman Shale, a PHPA polymer system comprising 1% by weight potassium chloride was developed with 4 Ibm/bbl HS. It was required that salt concentrations not go over 6,000 mg/L for disposal purposes. Before introducing the HS, 100 feet of formation were drilled in an effort to compare the fluid's performance with and without the HS. The cuttings had more mud qualities and were softer and "mushier" prior to treatment. The polymer system was then supplemented with 4 Ibm/bbl HS. When the HS was first applied, viscosity and fluid loss increased as predicted. Fluid characteristics started to stabilize as the concentration reached the required level, and drill cutting integrity significantly improved from an uncertain mass to well-defined cuttings that could pass through shaker screens with ease. There was no balling of the bit or bottom hole assembly during the shale section drilling, and the hole conditions held steady.[3]

At 4.0–4.5 Ibm/bbl, the HS content was sustained throughout the Waltman Formation. The well was sunk to total depth 12 days faster than any other comparable well done in the vicinity by removing the problematic Waltman Formation issues. Applying the HS during this trial supported the finding that pretreating the drilling fluid with HS before drilling a reactive formation is essential for the HS to function at its best. In addition, the comparatively stable characteristics (without requiring excessive water dilution) suggested using the HS as a rheology-control agent to reduce viscosity issues while drilling hydratable shale rather than only as a shale-stabilizing additive. The HS should help improve the effectiveness of solids-control equipment and lessen the need for dilution by reducing the dispersion propensity of drill cuttings. As a result, less mud would be used, and mud-treating expenses would go down. An excellent chance to confirm or refute the use of the HS to lower mud consumption needs was presented by a North Sea field testing. In this region of the North Sea, basic seawater/polyanionic cellulose (PAC) fluid is typically drilled at 12-in. intervals. To reduce the amount of reactive shale dispersion, low pH (pH < 9.0) and mud alkalinity levels are maintained. Shale samples obtained from offset wells were subjected to dispersion experiments before to field testing. In order to drill this interval, the base or standard fluid was compared to the same formulation that included 4 lbm/bbl HS.[2]

To rule out the chance of experimental mistake, two experiments were run. Compared to the untreated sample, the fluid containing the HS had a roughly 50% higher shale recovery. After that, the mud system was prepared using the HS (4 lbm/bbl) before the $12\frac{1}{4}$-inch interval was drilled. The additions (and losses) of water and mud treatment were thoroughly observed in order to calculate the interval's fluid consumption or dilution rate. According to the well data, the dilution rate for this period (using the HS) was 1.5 barrels per foot of drilled hole. The standard fluids typically utilized to drill this segment of hole had an average dilution rate of 1.8–2.0 barrels per foot of hole, according to an analysis of offset well logs.[2]

To ascertain the variety of fluid types that might profit from the usage of HS, more field testing of the device was started. In one such study, HS was used in an offshore Gulf of Mexico seawater/low lime/lignosulfonate fluid. Using lime and caustic soda, the pH range of this fluid was maintained between 11.0 and 11.5. The system was enhanced with HS before the test interval was drilled. Bit balling and bottom hole assembly were encountered during the "gumbo" shale drilling process. The comparison of mud treatment costs to maintain stable mud characteristics showed that they were less than offset wells, even if the addition of the HS did not seem to be successful in avoiding bit balling as was noted in prior testing (using lower pH range polymer fluids). According to this test, the HS cannot properly be categorized as a bit balling preventer because it seems to function best at lower pH levels (pH 9.5) when used as a shale control additive. Even at higher pH levels, the HS does seem to help stabilize the mud characteristics in fluids polluted by reactive shale cuttings.[2]

9.3 APPLICATION OF AMINO ACIDS AS CORROSION INHIBITOR

Description: The primary approach for preventing metals from corroding is the use of inhibitors, and several researchers are studying this issue. It is common to find new inhibitors. In theory, inhibitors work by interacting with the metal surface by adsorption through donor atoms, π-orbitals, electron density, and the molecule's electrical structure to prevent corrosion. Specifically, the $-NH_2$ and $-SH$ groups in inhibitor structures frequently serve as reaction hubs when the inhibitor is being adsorbed. By doing this, the inhibitors cover the metal surfaces with an adsorbed protective coating that shields them from the corrosive acid solution. Common corrosion inhibitors that pose a harm to the environment include chromates, nitrite, and aromatic heterocyclic chemicals. It is essential that these inhibitors be swapped out by new, environmentally friendly materials. Amino acids are entirely soluble in aqueous solutions, fully safe, biodegradable, and reasonably priced. The ability of certain amino acids to prevent metal from corroding has been extensively studied.[4]

Amino acids contain heteroatoms (e.g., S, N, and O) and eventual conjugated π-electrons system on their molecular structures. This attribute has attracted the attention of researchers to explore their potential abilities to act as green corrosion inhibitors. While an amino acid can effectively prevent corrosion on a certain metal under some circumstances, under other operating conditions it might actually hasten the corrosion process. The current focus of study is on using newly synthesized derivatives of amino acids as corrosion inhibitors, which have greater corrosion inhibition efficiency than their equivalent simpler/original molecules.[5]

9.3.1 Challenges

Traditional methods of corrosion prevention, such as coatings, inhibitors, and cathodic protection, have limitations in terms of effectiveness, environmental impact, and cost. Moreover, corrosion in certain environments, such as high-temperature and high-pressure conditions encountered in oil and gas production, presents unique challenges that require innovative solutions. Corrosion inhibition is a complicated process, and the effectiveness of inhibitors is dependent on a number of variables and the interaction of different competing effects. Immersion time can play a decisive role in the prevention of corrosion ability.[6]

9.3.2 Case Study 2: Amino Acid-Based Corrosion Inhibitor in Offshore Oil Production

One notable case study involves the application of an amino acid-based corrosion inhibitor in offshore oil production facilities. In this case, the corrosive environment of offshore platforms, characterized by exposure to seawater, high salinity, and aggressive chemicals, necessitated effective corrosion control measures to ensure the integrity and longevity of equipment and infrastructure.

9.3.3 Challenges Faced

The harsh operating conditions of offshore environments presented significant challenges for corrosion prevention, including:

1. **High Corrosivity of Seawater:** Seawater contains chloride ions, dissolved oxygen, and other corrosive species that accelerate metal degradation, posing a constant threat to offshore infrastructure.
2. **Temperature and Pressure Variations:** Offshore environments are subject to fluctuations in temperature and pressure, which can exacerbate corrosion rates and complicate corrosion management strategies.
3. **Safety and Environmental Concerns:** Corrosion-related failures in offshore facilities can have serious safety implications and environmental consequences, necessitating robust corrosion prevention measures.

To address these challenges, an amino acid-based corrosion inhibitor was introduced into the offshore production system. The corrosion inhibitor formulation was designed to be environmentally friendly, biodegradable, and effective in inhibiting both general and localized corrosion on metal surfaces.

The application of the amino acid-based corrosion inhibitor resulted in:

1. **Reduced Corrosion Rates:** The inhibitor effectively reduced corrosion rates on metal surfaces, prolonging the service life of equipment and infrastructure.
2. **Improved Asset Integrity:** Enhanced corrosion protection contributed to improved asset integrity and operational reliability, reducing the risk of unplanned downtime and maintenance costs.
3. **Environmental Compliance:** The environmentally friendly nature of the inhibitor formulation ensured compliance with regulatory requirements and minimized the impact on marine ecosystems.

9.4 APPLICATION OF AMINO ACID INTEGRATION IN HYDRAULIC FRACTURING FLUID SYSTEMS

Description: Hydraulic fracturing, a technique used to extract oil and natural gas from unconventional reservoirs, relies on the injection of fracturing fluids into the subsurface to create fractures and enhance hydrocarbon recovery. However, the efficiency and environmental impact of hydraulic fracturing operations can be influenced by the composition of the fracturing fluid. Traditional fracturing fluids often contain chemicals that may pose environmental and health concerns, leading to the exploration of alternative additives, such as amino acids, to improve fluid performance and sustainability.[7]

9.4.1 Challenges

Hydraulic fracturing operations face several challenges related to fluid performance, environmental impact, and regulatory compliance:

1. **Friction Reduction:** Friction between the fluid and the wellbore can limit the effectiveness of hydraulic fracturing operations, reducing fracture propagation and hydrocarbon recovery.

2. **Fluid Compatibility:** The composition of fracturing fluids must be compatible with reservoir conditions to prevent formation damage and ensure optimal well performance.
3. **Environmental Concerns:** Chemical additives used in fracturing fluids may pose environmental risks, including groundwater contamination and ecosystem disruption.

Solution with Amino Acids: Amino acids offer several potential benefits for hydraulic fracturing fluid systems:

1. **Friction Reduction:** Amino acids can act as friction reducers, reducing the energy required to pump fracturing fluids into the subsurface and improving fluid flow characteristics.
2. **Biodegradability:** Amino acids are biodegradable, minimizing their environmental impact and reducing the risk of long-term contamination.
3. **Compatibility:** Amino acids are compatible with reservoir conditions, allowing for effective fluid performance without compromising well integrity.

9.4.2 Case Study 3: Amino Acid-Based Friction Reducer in Hydraulic Fracturing

One notable case study involves the integration of an amino acid-based friction reducer in hydraulic fracturing fluid systems in the Permian Basin, a prolific oil and gas producing region in the United States. The Permian Basin, located in southeastern New Mexico and west Texas, is treated as the largest province of oil and gas production in the United States. The majority of hydrocarbon production from the Permian Basin is from the Wolfcamp shaly formation. Drilling horizontal and extended reached wells is continuously increasing day after day in the Permian Basin. Oil-based mud (OBM) is used to drill these horizontal wells for enhancing shale inhibition, reducing torque and drag, and supporting thermal stability of mud rheology. Due to environmental regulation for limiting use of OBM, the petroleum industry has tried to develop WBF that approaches the performance advantages of OBM.[8]

In the Permian Basin, hydraulic fracturing operations encounter challenges related to high friction pressures, which can limit fracture propagation and reduce hydrocarbon recovery rates. Traditional friction reducers used in fracturing fluids may contain toxic or environmentally harmful chemicals, leading to concerns about their long-term impact on groundwater quality and ecosystem health. To address these challenges, a novel amino acid-based friction reducer was introduced into the hydraulic fracturing fluid system. The friction reducer formulation was designed to be environmentally friendly, biodegradable, and effective in reducing friction pressures during hydraulic fracturing operations.[9]

The application of the amino acid-based friction reducer resulted in:

1. **Improved Fracture Propagation:** The friction reducer effectively reduced friction pressures, allowing for improved fracture propagation and increased hydrocarbon recovery rates.

2. **Environmental Compliance:** The environmentally friendly nature of the amino acid-based friction reducer ensured compliance with regulatory requirements and minimized the risk of environmental contamination.
3. **Cost Savings:** The use of the amino acid-based friction reducer led to cost savings by reducing the amount of chemical additives required in the fracturing fluid system and improving overall operational efficiency.

9.5 CONCLUSION

The three case studies demonstrate how amino acids may be successfully integrated into difficult industrial applications, demonstrating their adaptability and efficiency in solving important problems. Regarding corrosion protection, amino acid-based inhibitors have shown encouraging results in reducing the deterioration of metal in corrosive conditions, providing a long-term substitute for conventional corrosion control techniques. Amphoteric compounds, like amino acids, inhibit clay hydration in aqueous solutions that are either saline or nonsaline. The HS may be added to mud systems without negatively affecting the mud's characteristics. In fresh muds containing the HS, higher quantities of fluid-loss-control additives could be necessary. It is necessary to do further research and analysis on how these amphoteric compounds interact chemically with clays. These substances are thought to be further improved as shale-controlling agents, especially in distributed mud systems with higher pH levels.[2] Amino acid-based additives have also shown promise in hydraulic fracturing operations for improving fluid performance, lessening environmental impact, and maintaining regulatory compliance.

These case studies highlight how amino acids have the ability to transform a range of industrial processes, from hydraulic fracturing to corrosion prevention, by providing sustainable, affordable, and eco-friendly solutions. Industry participants may overcome challenges pertaining to performance, environmental impact, and regulatory compliance by utilizing the special qualities of amino acids, which will eventually result in safer, more effective, and more sustainable operations.

REFERENCES

[1] Quainoo AK, Negash BM, Bavoh CB, Idris A. Natural amino acids as potential swelling and dispersion inhibitors for montmorillonite-rich shale formations, *J Pet Sci Eng.* 2021;196:1–2.

[2] Bruton JR, Mclaurine HC. *Modified Poly-Amino Acid Hydration Suppressant Proves Successful in Controlling Reactive Shales* (1993).

[3] Alshibli FH, Alhaleem AA. Formulating Inhibited Fluids for Stable Drilling Operations into Tanuma and Zubair Shales, Zubair Oilfield, Southern Iraq, *Iraq Geol J.* 2022;55:117.

[4] Jabbar AH, Kamona SMH, Abbood SK., Hussein TK, Al-Saidi DN, Hameed SM, Rashid RAK, Abbas HA, Kadhim MM. The effective and sustainable application of a green amino acid-based corrosion Inhibitor for Cu metal. *Chem Phys Impact.* 2023;7:1.

[5] Ekeke IC, Efe S, Nwadire FC. A review of amino acids used as corrosion inhibitors on iron metal/alloys in aggressive environments, *Mater Prot.* 2022;63:319.

[6] Ibrahimi BEl, Jmiai A, Bazzi L, Issami SEl. Amino acids and their derivatives as corrosion inhibitors for metals and alloys,*Arab J Chem.* 2020;13:746.

[7] Fjær E, Holt RM, Horsrud P, Raaen AM, Risnes R. in *Developments in Petroleum Science,* 153–154 (2021).
[8] Metwally M, Nguyen T, Wiggins H, Saasen A, Gipson M, Yoo H. Evaluations of Polyacrylamide Water-Based Drilling Fluids for Horizontal Drilling in the Shaly Wolfcamp Formation, *SPE J.* 2023;28:1744.
[9] Rodriguez J, Heo J, Kim KH. The impact of hydraulic fracturing on groundwater quality in the permian Basin, West Texas, USA, *Water (Switzerland).* 2020;12:8.

Index

Note: **Bold** page numbers refer to tables and *italic* page numbers refer to figures.

For Product Safety Concerns and Information please contact our EU representative GPSR@taylorandfrancis.com Taylor & Francis Verlag GmbH, Kaufingerstraße 24, 80331 München, Germany